James Morton

The System of Calculating Diameter, Circumference, Area and Squaring the Circle

James Morton

The System of Calculating Diameter, Circumference, Area and Squaring the Circle

ISBN/EAN: 9783337404536

Printed in Europe, USA, Canada, Australia, Japan

Cover: Foto ©berggeist007 / pixelio.de

More available books at **www.hansebooks.com**

THE

SYSTEM OF CALCULATING

DIAMETER, CIRCUMFERENCE, AREA,

AND

SQUARING THE CIRCLE.

TOGETHER WITH

INTEREST AND MISCELLANEOUS TABLES, AND
OTHER INFORMATION.

BY

JAMES MORTON.

PREFACE.

It is not the purpose of the writer to introduce to the public any new principle, but the result of laborious calculations culminating in the final elucidation of facts.

The Interest Table is calculated on the usual custom adopted by bankers — of 30 days to the month.

The other tables are collated from the best authorities.

CONTENTS.

1 *   v

DIAMETER, CIRCUMFERENCE, AREA,

AND

SQUARING THE CIRCLE.

The Circle, and the Measurement of Angles.

Definitions.

A CIRCLE is a plane figure bounded by a curved line, every part of which is equidistant from a point within called the centre, and contains as great an area within the same outline or perimeter as any other form.

The line, A B, drawn through the centre is called a diameter.

The straight line drawn from the centre, D, to the circumference, C, is called a radius or semidiameter.

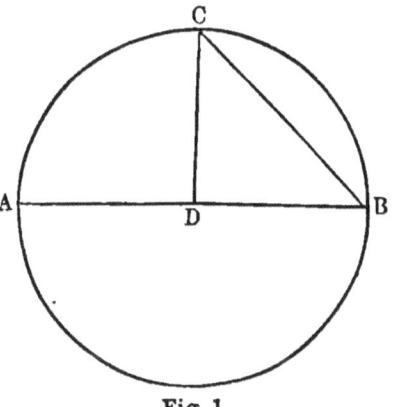

Fig. 1.

Any part of a circumference is called an arc. A C being an arc of the one-fourth of the circumference.

The straight line, C B, is called a chord.

A diameter is greater than any other chord.

9

Every diameter divides the circle and its circumference each into two equal parts.

To Inscribe an Equilateral Triangle in a given Circle.

To inscribe an equilateral triangle in a circle, divide the circle into three equal parts; draw the chords A C, A B, C B.

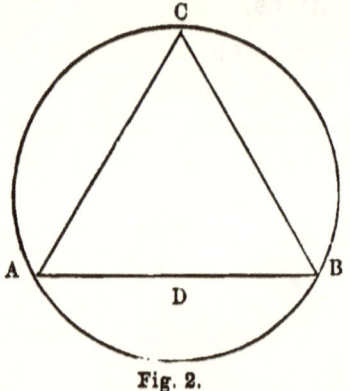

Fig. 2.

The area of this triangle is equal to half the product of its base and altitude.

Let A B C be a triangle, and B D perpendicular to the base; then will its area be equal to one-half of A C × B D.

For draw C E parallel to A B, and B E parallel to A C, completing the parallelogram A E. Then the triangle, A B C, is half the parallelogram A B C E, which has the same base, A C, and the same altitude, B D; but the

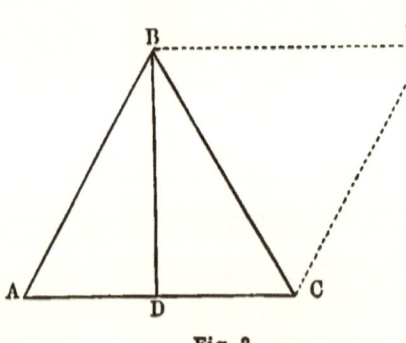

Fig. 3.

area of the parallelogram is equal to A C × B D; hence, that of the triangle must be ½ A C × B D or A C × ½ B D.

Two triangles of equal altitude are to each other as their bases, and two triangles of equal bases are to each other as their altitudes; and triangles generally are to each other as the products of their bases and altitudes.

The square of the one side of an equilateral triangle inscribed in a circle is equal to three-fourths the area of a square described without the circle and tangent thereunto. Therefore, by squaring the diameter and extracting the square root of three-fourths the product, you have the one side of an equilateral triangle drawn in a circle.

EXAMPLE.— A circle the diameter of two hundred square thirds equal to 115.47 + feet and the area of the circumscribed square being thirteen thousand three hundred and thirty-three and a third square feet.

$$
\begin{array}{r}
200 \\
\underline{200} \\
\tfrac{1}{3})\overline{40000} \\
\underline{13333\tfrac{1}{3}} \\
6666\tfrac{2}{3} \\
\underline{3333\tfrac{1}{3}} \\
\overline{10000} = \tfrac{3}{4} \text{ the area.}
\end{array}
$$

The square root of (10000) ten thousand feet equals one hundred feet, the side of the equilateral triangle.

Or, multiply the diameter by the decimal $.86\tfrac{6}{10}$.

For the area of the triangle, multiply the square of the side of the triangle by the decimal .4330.

To Inscribe a Square in a Given Circle.

To inscribe a square in a circle, draw two diameters, A B, C D, intersecting each other at right angles; join their extremities, A C B D.

A square inscribed inside of a circle is half the size of a square described outside of a circle.

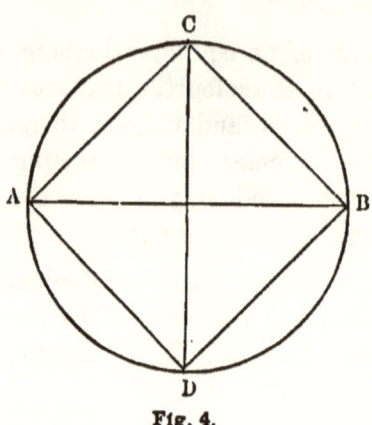

Fig. 4.

For the side of an inscribed square, multiply the diameter by the decimal .70711.

Let the diameter, A B, be 128 feet; then 128 × 70711 = $90\frac{510}{1000}$. Near enough for all practical purposes.

You cannot extract the square root from 128, an imperfect power, but you may approximate it. 64 is a perfect power and a square number, as $8 \times 8 = 64$; but when you take a prime number, you must reduce it to a fractional number to extract the square root, as 1 reduced to $144 = 12 \times 12$, the root being $\frac{12}{144}$; two equivalent to $\frac{256}{128}$, the square of which is $\frac{16}{128}$, and three, $\frac{9}{2}$.

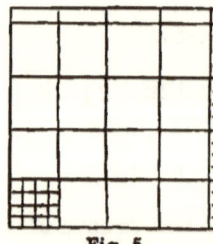

Fig. 5.

This square, $4\frac{1}{4} \times 4\frac{1}{4}$ feet, equals $18\frac{1}{16}$ square feet, the square root of which is seventeen sixteenths, $4.25 \times 4.25 = 18\frac{625}{10000}$, $4\frac{1}{4} \times 4\frac{1}{4} = 18\frac{1}{16}$ square feet, reduced to sixteenths, equal 17 square sixteenths on a side, $17 \times 17 = 289$. $289 \div 16 = 18\frac{1}{16}$.

NOTE.—A square number cannot have more places of figures than double the places of the root, but sometimes less; for instance 10, the square of which is 100, one less than double the places of the root.

When the number of places in a given sum is an odd number, the left hand period will contain one figure, as 1.44; but the root will be composed of as many figures as there are periods, 12 being the square root of 144.

Explanation of the Process of Extracting the Square Root.— First, point off the given number into periods of two figures each, by putting a dot between every two figures to the left, and also to the right, when there are decimals.

Secondly, find the greatest square in the left hand period, and write its root in the quotient. Subtract the square of this root from the left hand period, and to the remainder bring down the next period for a dividend.

Thirdly, double the root already found for a divisor; ascertain how many times the divisor is contained in the dividend, excepting the right hand figure, and place the result in the root, also at the right hand of the divisor and under the same, and multiply the divisor by this figure. Subtract the product from the dividend, and to the remainder bring down the next period for a new dividend, and so continue until all the periods are brought down.

If any dividend shall be too small to contain the divisor, place a cipher in the root, also to the right of the divisor, and bring down the next period to the right hand of the dividend, and proceed in the work.

Take eighteen square feet and extract the square root by decimals, thus:

```
          18 ( 4.24264 +
          16
  .
 82      200
  2      164          .
 ___     ___
 844     3600      84846     543600
   4     3376          6     509076
 ____    _____      _____   _____
 8482    22400      848524   3452400
    2    16964           4   3394096
 _____   _____     _____   _____
 84846   543600     848528   58304
  2
```

You may proceed in this manner, *ad infinitum*, without reaching a finality; but in the following process you may always accomplish a final result.

$$18\,(\,4\tfrac{2}{8}$$
$$16$$
$$\overline{\;8\;\;2}$$

$$4\tfrac{2}{8}$$
$$4\tfrac{2}{8}$$
$$\overline{16}$$
$$1$$
$$1$$
$$\overline{18}$$

This is not $4\tfrac{2}{8} \times 4\tfrac{2}{8}$, because that would make the same result as the previous example, but $\tfrac{2}{8}$ of 4, and not of the fraction; its equivalent being more than four and twenty-four hundredths, and less than twenty-five hundredths, about $4.24\tfrac{47488}{179888}$.

The square root of 288 by this process.

$$288\,(\,16\tfrac{32}{32}$$
$$1$$
$$\overline{\;\;188}$$

$$16\tfrac{32}{32}$$
$$16\tfrac{32}{32}$$
$$\overline{\;96}$$

$$26 \qquad 188$$
$$\;6 \qquad 156$$
$$\overline{32} \qquad \overline{\;32}$$

$$96$$
$$16$$
$$16$$
$$16$$
$$\overline{288}$$

This is sixteen times sixteen, and thirty-two thirty-seconds of sixteen — twice for the four sides of the square, thirty-two representing two sides. That is, the square of sixteen equals two hundred and fifty-six and the remainder; viz., thirty-two is proportionately distributed by this process, and, by the other process, the square root of the same is as follows:

$$16.970562748477140583562+.$$

To Find the Approximate Square Roots of Surds or Prime Numbers, and the Correct Area of their Squares.

Take, for example, one hundred feet square, equal to ten thousand square feet; the side of this square being one hundred feet, and the half of this square being five thousand square feet, and a prime number can only be squared in the following manner:

$$50.00\ (\ 70\tfrac{100}{140}$$
$$49$$
$$\overline{140\quad 100}$$

$$70$$
$$70$$
$$\overline{4900}$$
$$50$$
$$\overline{50}$$
$$\overline{50.00}$$

$$70$$
$$100$$
$$140\ )\ \overline{7000}\ (\ 50$$
$$700$$
$$\overline{0}$$

Reducing the side to fractions of one hundred and forty, and squaring the same, then dividing by the square of one hundred and forty, you have as follows:

$$140$$
$$140$$
$$\overline{5600}$$
$$140$$
$$\overline{19600}$$

$$140$$
$$70\tfrac{100}{110}$$
$$\overline{9800}$$
$$100$$
$$\overline{9900}$$
$$9900$$
$$\overline{8910000}$$
$$89100$$
$$\overline{19600\ )\ 98010000\ (\ 5000}$$
$$98000$$
$$\overline{10000}$$

This remainder, 10000, is the square of the numerator of the fraction $\tfrac{100}{140}$, viz., $100 \times 100 = 10000$; the remainder will always correspond to the square of the numerator of the fraction, and the product will always be plus the square of the numerator. Again, reduce 5000 to square $140 = 98000000$, thus:

140	140	140
140	70	10
5600	9800	1400
140	9800	1400
19600	7840000	560000
5000	88200	1400
98000000	96040000	1960000
	1960000	
	98000000	

The remainder, or fraction, is × 10, its square root. As proof of the above calculation, take the square 100.

140	140
140	100
5600	14000
140	14000
19600	56000000
10000	14000
2)196000000	196000000
98000000	

		98.00.00.00 (9899$\frac{9799}{19798}$		9899
		81		9799
188	1700	9899		89091
8	1504	9899		89091
1969	19600	89091		69293
9	17721	89091		89001
19789	187900	79192	19798)97000301 (4899$\frac{1}{2}$
9	178101	89091	79192
19798	9799	4899$\frac{1}{2}$	178083
		4899$\frac{1}{2}$	158384
		98000000	196990
			178182
			188081

In this manner can be calculated the area of the diagonal square, and the correct amount be ascertained; but the exact side of the diagonal square is only approximated, the side being less than $70\frac{100}{140}$, that is, 70×70 and $\frac{100}{140}$ of 70 on the two angular sides, and in this manner the square root of 2 may be also approximated, viz.:

$$1.414213\tfrac{7}{14}\tfrac{9}{14}\tfrac{5}{14}\tfrac{3}{2}\tfrac{1}{1}\tfrac{5}{3}\tfrac{\frac{1}{2}}{}$$
$$1.414213\tfrac{7}{14}\tfrac{9}{14}\tfrac{5}{14}\tfrac{3}{2}\tfrac{1}{1}\tfrac{5}{3}\tfrac{\frac{1}{2}}{}$$

4242639
1414213
2828426
5656852
1414213
5656852
1414213

$$7953315\tfrac{1}{2}$$

A complex fraction of

$$7953315\tfrac{1}{2}$$

$$1.414213\dfrac{7953315\frac{328\frac{1}{2}}{900}}{1414213}.$$

2.000000000000

Let the diagonal be 16 feet square. $16 \times 16 = 256$ square feet. Double this for the square on the diagonal.

$$256 \times 2 = 5.12\ (\ 22\tfrac{28}{44}$$

$$\begin{array}{cc}
 & 4 \\
\hline
42 & 112 \\
2 & 84 \\
\hline
44 & 28 \\
\end{array}$$

This, reduced to square forty-fourths, is $996 \times 996 = 992016$
Subtract the square of the numerator, $28 \times 28 =$ 784

991232

Divide this by the square of $44 = 512$.

512 reduced to 2048 $= \frac{1048576}{2048}$.

2 *   B

The square root of this is $\frac{1024}{2048}$, being equivalent to 22 feet, 7 inches, $\frac{524}{100}$ of an inch on a side of the square.

The Square described on the Diagonal of a Given Square is Equivalent to Double the Given Square.

Let E H F G be a square described on E H, and A B C D a square described on the diagonal, E F; the triangle, E

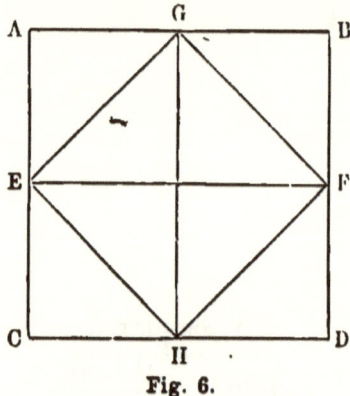

Fig. 6.

H F, being right-angled and isosceles, you have $\overline{EF}^2 = \overline{EH}^2 + \overline{HF}^2 = 2\overline{EH}^2$. Hence, the area of a square is equivalent to double the area of the diagonal.

When four magnitudes are in proportion, the product of the two extremes is equal to the product of the two means.

If four magnitudes are in proportion, they will be in proportion if taken inversely.

Equimultiples of any two magnitudes have the same ratio as the magnitudes themselves.

If there be four proportionate magnitudes, and four other proportionate magnitudes having the antecedents the same on both, then consequents will be proportional.

As the sum of the one side of a square is to the sum of the three sides so is the square of half the sum of the one side to three-quarters the area of the square; for example, let the diagonal be 300 feet on a side.

$$300 : 900 :: 150^2 : 67500 ;$$

and if 67500 equals three-fourths the area of the square,

90000 equals the area. Changing the places of the members of the equation, we have

$$67500 : 22500 :: 900 : 300.$$

And 180000 square feet being the area of double the diagonal or the square on the diagonal, we have

$$135000 : 45000 :: 1272\tfrac{336}{424} : 424\tfrac{112}{424};$$

consequently, the side of a square is to the side of a diagonal

as 2715288960000000 is to 19199992364969968½

nearly, because there is no common measure for the two, unless it is the square of the side, as follows:

The square of 300 = 90.000.
The square of $424\tfrac{112}{424}$ = 180.000.

Two Sides of a Right-Angled Triangle Given to Find the Other.

Let A B C D be a square of 100 feet on a side.

The sides, A K, I K, of the right-angled triangle, A I K, are each equal, and twenty-five feet each; and the square root of the sum of the two sides equals the hypothenuse, A I, as follows:

```
        25
        25
       ———
       125
        50
       ———
       625
       625
      ————
      1250 ( 35 25/70
         9
 65    ———
  5    350
 ——    325
 70    ———
        25
```

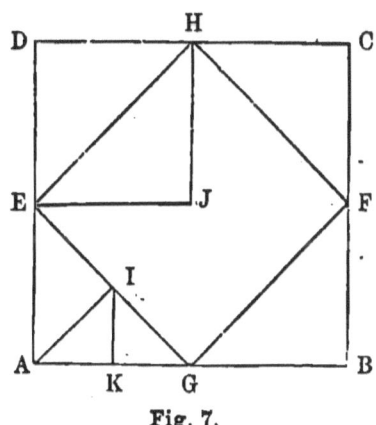

Fig. 7.

The sides, A I, E I, of the right-angled triangle, E I A, are each $35\frac{25}{70}$, and the square root of the sum of the two sides squared equals the side, A E, viz.:

	$35\frac{25}{70}$	35
	$35\frac{25}{70}$	25
	175	175
50	105	70
50	12 — 35	70) 875 (12
2500	12 — 35	70
	1250	175
	1250	140
	2500 (50	35
	25	
	00	

The sides, E J and H J, each fifty feet, being squared, the square root of the sum of the two is equal to the diagonal E H; and the side of the square, E G F H 70$\frac{100}{140}$, and the two sides E G and F G 70$\frac{100}{140}$ being squared, the square root of the sum of the two equals 100, the side of the square A B C D, this fraction $\frac{100}{140}$ is a little over, it being about $\frac{98}{140}$ or $\frac{100}{140}$ of 100.

NOTE. — There are some numbers the sum of whose squares being a perfect square leave no remainder or fraction. Such are 3 and 4, the sum of whose squares is 25, and the square root 5; therefore the hypothenuse is 5; and if those numbers be multiplied by other numbers, each of the same, the products will be the sides of true right-angled triangles. Multiplying them by 2 gives 6, 8, and 10, which are used by builders in laying out corners.

EXAMPLE.—Suppose G B to be 4 yards in length, and B F to be 3 yards in height; then the square of G B is 16 yards, and the square of B F is 9 yards, and the sum of their squares is 25 yards.

The square root of 25 yards is 5 yards, which is the length of the hypothenuse.

A Regular Polygon is One which is both Equilateral and Equiangular.

A regular polygon may have any number of sides; the equilateral triangle is one of three sides; the square is one of four.

To inscribe in a circle a regular hexagon, let the circle be one hundred and twenty-eight feet diameter. Beginning at a point, B, apply the radius, B O, six times as a chord to the circumference, and you will form the regular hexagon, B C D E F A. Draw O H perpendicular to one of its sides; the area of the polygon is equal to

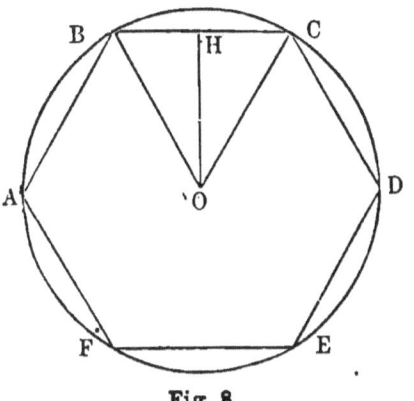

Fig. 8.

$\frac{1}{2}$ O H multiplied by the perimeter. Now let the arcs which are subtended by the sides of the polygon be bisected, and new polygons formed; the limit of the perimeter is the circumference of the circle.

The area of the right-angled triangle, B O C, is equal to a rectangle $27\frac{11}{100} \times 64 = 1773\frac{6}{10}$, being the one-sixth of the area of the polygon, which is equal to a rectangle $96 \times 110\frac{85}{100} = 10641\frac{6}{10}$ feet.

To find the area of a triangle when the base and altitude are given, multiply the base by the altitude, and take half the product, or multiply one of these dimensions by half the other.

The circumferences of circles are to each other as their radii, and the areas are to each other as the squares of their radii.

The circumferences of circles are to each other as their diameters, and their areas are to each other as the squares of their diameters.

The circumference of the circle is the limit of all inscribed polygons; in fact, the circle is but a regular polygon of an infinite number of sides.

If a regular hexagon be inscribed in a circle, its side will be equal to the radius.

The Area of a Circle is Equal to the Product of Half the Radius by the Circumference.

Let A C D E be a circle whose centre is O and radius O A; then will area O A $= \frac{1}{2}$ Q A × cir. O A.

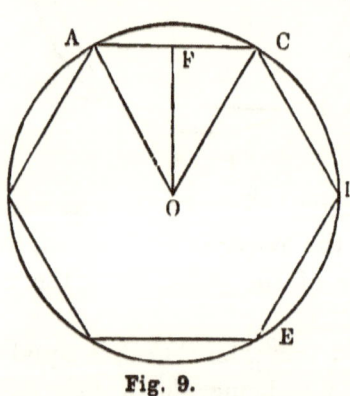

Fig. 9.

For inscribe in the circle any regular polygon, and draw O F perpendicular to one of its sides. The area of the polygon is equal to $\frac{1}{2}$ O F multiplied by the perimeter. Now let the arcs which are subtended by the sides of the polygon be bisected, and new polygons formed, as before. The limit of the perimeter is the circumference of the circle; the limit of the apothegm is the radius O A, and the limit of the area of the polygon is the area of the circle passing to the limit; the expression for the area is area $O A = \frac{1}{2} O A \times cir. O A$;

consequently, the area of a circle is equal to the product of half the radius by the circumference.

The problem of the quadrature of the circle, as it is called, consists in finding a square equivalent in surface to a circle, the radius of which is known. Now, it is proved that a circle is equivalent to the rectangle contained by its circumference and half its radius; and this rectangle may be changed into an equivalent square by finding a mean proportional between its length and its breadth. To square the circle, therefore, is to find the circumference when the radius is given; and for effecting this it is enough to know the ratio of the diameter to the circumference.

Archimedes showed that the ratio of the diameter to the circumference is included between $3\frac{10}{70}$ and $3\frac{10}{71}$, hence $3\frac{1}{7}$ or $\frac{22}{7}$.

Metius, for the same quantity, found the much more accurate value, $\frac{355}{113}$.

_____

Proposition, Theorem, and Calculation of the Limit of the Circle.

Let A B C be a circle, of which the centre is D and the diameter, A C, 128 feet.

In the circle, A B'C, apply the straight line, E F, equal to the radius, D C; draw also D F, D E, and B F. It is evident that the arc E B F is one-sixth of the circumference.

The equilateral triangle, E D F, is the one-sixth part of a six-sided polygon, the perimeter of which is three times the diameter of the circle, viz., 384 feet.

The line G F, or half of E F, is 32 feet, but the line B F is $33.1288377731226657\frac{6}{10}$; and B F is the side of a twelve-sided polygon drawn inside the circle: it is evident that as you increase the sides of the polygon you approach the circumference of the circle. Continuing the increase to a polygon of three million one hundred and forty-five thousand seven hundred and twenty-eight sides (3,145,728), you proceed as follows: the hypothenuse, D F, of right-angled triangle

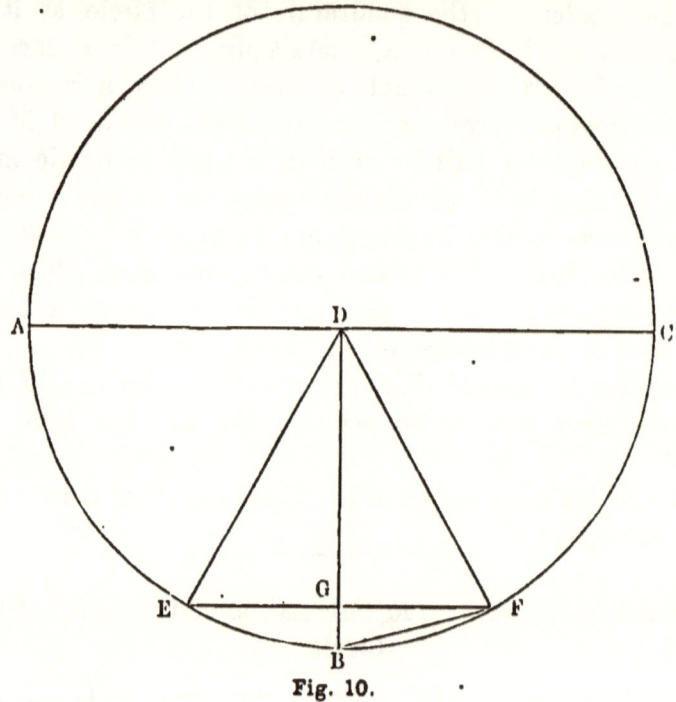

Fig. 10.

D G F, being sixty-four feet, and the base, G F, thirty-two, the square root of the difference of the squares of the same is equal to the perpendicular D G, 55.425625842204073; and subtracting D G from D B gives the base line G B of right-angled triangle G B F, and the square root of the sum of the squares of G F and G B gives the dimensions of the line B F, therefore the half of B F is the base line of a right-angled triangle of a twenty-four-sided polygon, and so continue until you reach a polygon of 3145728 sides.

Calculation of Right-angled Triangle in Polygon of 12 Sides.

EXAMPLE.

30.72 (55.425625842204073
25

32	. 105	572
32	5	525
64	1104	4700
96	4	4416
1024	11082	28400
	2	22164
	110845	623600
	5	554225
	1108506	6937500
	6	6651036
64	11085122	28646400
64	2	22170244
256	110851245	647615600
384	5	554256225
4096	1108512508	9335937500
1024	8	8868100064
3072	11085125164	46783743600
	4	44340500656
	110851251682	244324294400
	2	221702503364
	1108512516842	2262179103600
	2	2217025033684
	110851251684404	451540699160000
	4	443405006737616
	11085125168440807	81356924223840000
	7	77595876179085649
	11085125168440814	37610480447543510 0
	3	3325537550532244 29

3

64.

55.425625842204073

8.574374157795927

8.574374157795927
8.574374157795927
_____
60020619104571489
17148748315591854
77169367420163343
42871870788979635
77169367420163343
60020619104571489
60020619104571489
42871870788979635
8574374157795927
34297496631183708
60020619104571489
25723122473387781
34297496631183708
60020619104571489
42871870788979635
68594993262367416
_____

73.519892197878612448950577789329
1024.
_____
1097.519892197878612448950577789329

10.97.51.98.92.19.78.78.61.24.48.95.05.77.78.93.29 (33.1288377731
$\qquad$ 9 $\qquad\qquad\qquad\qquad$ [22657$\frac{6}{10}$

63	197
3	189

661	851
1	661

6622	19098
2	13244

2) 33.128837773122657.6

16.564418886561328.8

66248	585492
8	529984

662568	5550819
8	5300544

6625763	25027578
3	19877289

66257667	515028978
7	463803669

662576747	5122530961
7	4638037229

6625767547	48449373224
7	46380372829

66257675543	206900039548
3	198773026629

662576755461	812701291995
1	662576755461

6625767554622	15012453653405
2	13251535109244

66257675546242	176091854416177
2	132515351092484

662576755462446	4357650332369378
6	3975400532774676

6625767554624525	38218979959470293
5	33128837773222625

66257675546245307	509014218624766829
7	463803728823717149

Calculation of Right-angled Triangle in Polygon of 24 Sides.

EXAMPLE.

$$16.564418886561328_{16}^{8}$$
$$16.564418886561328_{16}^{8}$$

132515351092490624
33128837773122656
496932566596083984
16564418886561328
99386513319367968
82822094432806640
99386513319367968
132515351092490624
132515351092490624
132515351092490624
16564418886561328
66257675546245312
66257675546245312
99386513319367968
82822094432806640
99386513319367968
16564418886561328
$\qquad$ 13251535109249062
$\qquad$ 13251535109249062

274.37997304946965174826652362170S

64
64
———
256
384
———
4096.

2 74.37 99 73 04 94 69 65 17 48 26 65 23 62 17 08
————————————————————————————————
38.21.62.00.26.95.05.30.34.82.51.73.34.76.37.82.92 (61.819252882
36   [$500370\tfrac{3}{10}$

| 121 | 221 |
| 1 | 121 |

| 1228 | 10062 |
| 8 | 9824 |

| 12361 | 23800 |
| 1 | 12361 |

64.
61.819252882500370.3
2.180747117499629.7

| 123629 | 1143926 |
| 9 | 1112661 |

| 1236382 | 3126595 |
| 2 | 2472764 |

| 12363845 | 65383105 |
| 5 | 61819225 |

| 123638502 | 356388030 |
| 2 | 247277004 |

| 1236385048 | 10911102634 |
| 8 | 9891080384 |

| 12363850568 | 102002225082 |
| 8 | 98910804544 |

| 123638505762 | 309142053851 |
| 2 | 247277011524 |

| 1236385057645 | 6186504232773 |
| 5 | 6181925288225 |

| 1236385057650003 | 4578944548347637 |
| 3 | 3709155172950009 |

| 12363850576500067 | 86978937539762882 |
| 7 | 86546954035500409 |

| 12363850576500740 | 43198350426241392 |

3 *

$$2.180747117499629\tfrac{7}{10}$$
$$2.180747117499629\tfrac{7}{10}$$

19626724057496661
4361494234999258
13084482704997774
19626724057496661
19626724057496661
8722988469998516
15265229822497403
2180747117499629
2180747117499629
15265229822497403
8722988469998516
15265229822497403
17445976939997032
2180747117499629
4361494234999258
1526522982249740.3
1526522982249740.3

4755657990048204374403477903 7122

2) 16.707352604166603.7

8.353676302083301.8½

2 74.37 99 73 04 94 69 65 17 48 26 65 23 62 17 08
4.75 56 57 99 04 82 94 37 44 93 47 79 63 71 22
_____
2.79.13.56.31.03.99.52.59.54.93.20.13.03.25.88.30 (16.70735260416
1   [6603$\frac{7}{10}$

26	179
6	156
327	2313
7	2289
33407	245631
7	233849
334143	1178203
3	1002429
3341465	17577499
5	16707325
33414702	87017452
2	66829404
334147046	2018804859
6	2004882276
33414705204	139225835493
4	133658820816
334147052081	556701467720
1	334147052081
3341470520826	22255441563913
6	20048823124956
33414705208326	220661843895703
6	200488231249956
334147052083326	2017361264574725
6	2004882312499956
33414705208333203	124789520747698830
3	100244115624999609

Summary of Calculating the Angles of Polygons from 12 to 3145728.

Polygon of 12 Sides.

Base . . . . . . 32. feet,
Perpendicular 55.425625842204073,
Hypothenuse . . . . 64.

Base 8.574374157795927,
Perpendicular . . . 32.
Hypothenuse . . . . 33.128837773122657$\frac{6}{10}$.

Polygon of 24 Sides.

Base . . . . . . 16.564418886561328$\frac{8}{10}$,
Perpendicular . . . 61.819252882500370$\frac{3}{10}$,
Hypothenuse . . . . 64.

Base . . . . . . 2.180747117499629$\frac{7}{10}$,
Perpendicular ·. . . : 16.564418886561328$\frac{8}{10}$,
Hypothenuse . . . . 16.707352604166603$\frac{7}{10}$.

Polygon of 48 Sides.

Base 8.353676302083302,
Perpendicular . . . 63.452471127923866,
Hypothenuse . . . . 64.

Base 547528872076134,
Perpendicular . . . 8.353676302083302,
Hypothenuse . . . . 8.371600541458312.

Polygon of 96 Sides.

Base 4.185800270729156,
Perpendicular . . . 63.862971087270624,
Hypothenuse . . . . 64.

Base . . . . ·. . .137028912729376,
Perpendicular ·. . . . 4.185800270729156,
Hypothenuse . . . . 4.188042601187346.

Polygon of 192 Sides.

Base 2.094021300593673,
Perpendicular 63.965733598487401,
Hypothenuse . . . . 64.

Base 034266401512599,
Perpendicular 2.094021300593673,
Hypothenuse . . . . 2.094301648190307.

Polygon of 384 Sides.

Base 1.047150824095153½,
Perpendicular 63.991432826211954,
Hypothenuse . . . . 64.

Base 008567173788046,
Perpendicular 1.047150824095153½,
Hypothenuse . . . . 1.047185869303952.

Polygon of 768 Sides.

Base 523592934651976,
Perpendicular . . . . 63.997858170713670$\frac{6}{10}$,
Hypothenuse . . . . 64.

Base . . . . . . .002141829286329$\frac{4}{10}$,
Perpendicular 523592934651976,
Hypothenuse . . . . .523597315358052$\frac{8}{10}$.

Polygon of 1536 Sides.

Base . . . . . . .261798657679026$\frac{4}{10}$,
Perpendicular . . . . ₁ 63.999464540438441½,
Hypothenuse . . . . 64.

Base 000535459561558½,
Perpendicular . . . . . .261798657679026$\frac{4}{10}$,
Hypothenuse . . . . .261799205269004.

C

Polygon of 3072 Sides.

Base	.130899602634502,
Perpendicular	63.9998661349696$11\frac{6}{10}$,
Hypothenuse	64.

Base	.000133865030388$\frac{4}{10}$,
Perpendicular	.130899602634502,
Hypothenuse	.130899671083302$\frac{8}{10}$.

Polygon of 6144 Sides.

Base	.065449835541651$\frac{4}{10}$,
Perpendicular	63.999966533733652$\frac{9}{10}$,
Hypothenuse	64.

Base	.000033466266347$\frac{1}{10}$,
Perpendicular	.065449835541651$\frac{4}{10}$,
Hypothenuse	.065449844097753$\frac{1}{10}$.

Polygon of 12288 Sides.

Base	.032724922048876$\frac{1}{10}$,
Perpendicular	63.99999163343294$4\frac{6}{10}$,
Hypothenuse	64.

Base	.000008366567055$\frac{5}{10}$,
Perpendicular	.032724922048876$\frac{1}{10}$,
Hypothenuse	.032724923118388$\frac{8}{10}$.

Polygon of 24576 Sides.

Base	.016362461559194$\frac{4}{10}$,
Perpendicular	63.999997908358182$\frac{4}{10}$,
Hypothenuse	64.

Base	.000002091641817$\frac{6}{10}$,
Perpendicular	.016362461559194$\frac{4}{10}$,
Hypothenuse	.016362461692883.

Polygon of 49152 Sides.

Base 0081812308464411½,
Perpendicular . . . . 63.999999477089543$\frac{4}{10}$,
Hypothenuse . . . . 64.

Base . . . . . . .000000522910456$\frac{6}{10}$,
Perpendicular 0081812308464441,
Hypothenuse . . . . .0081812308631521$\frac{6}{10}$.

Polygon of 98304 Sides.

Base . . . . . . .0040906154315761$\frac{3}{10}$,
Perpendicular . . . . 63.9999998692723851$\frac{7}{10}$,
Hypothenuse . . . . 64.

Base . . . . . . .0000001307276141$\frac{3}{10}$,
Perpendicular . . . . .0040906154315761$\frac{3}{10}$,
Hypothenuse . . . . .0040906154336651$\frac{2}{10}$.

Polygon of 196608 Sides.

Base . . . . . . .0020453077168321$\frac{6}{10}$,
Perpendicular . . . . 63.9999999673180961$\frac{4}{10}$,
Hypothenuse . . . . 64.

Base . . . . . . .0000000326819031$\frac{6}{10}$,
Perpendicular . . . . .0020453077168321$\frac{6}{10}$,
Hypothenuse . . . . .0020453077170931$\frac{7}{10}$.

Polygon of 393216 Sides.

Base . . . . . . .0010226538585468$\frac{1}{10}$,
Perpendicular . . . . 63.9999999918295241$\frac{1}{10}$,
Hypothenuse . . . . 64.

Base . . . . . . .0000000081704751$\frac{9}{10}$,
Perpendicular . . . . .0010226538585468$\frac{1}{10}$,
Hypothenuse . . . . .0010226538585791$\frac{4}{10}$.

Polygon of 786432 Sides.

Base . . . . . $.000511326929289\tfrac{7}{10}$,
Perpendicular . . . 63.999999997957381,
Hypothenuse . . . $64.$

Base . . . . . $.000000002042619$,
Perpendicular . . . $.000511326929289\tfrac{7}{10}$,
Hypothenuse . . . $.000511326929293\tfrac{7}{10}$.

Polygon of 1572864 Sides.

Base . . . . . $.000255663464646\tfrac{84}{10}$,
Perpendicular . . . $63.9999999994893451\tfrac{2}{10}$,
Hypothenuse . . . $64.$

Base . . . . . $.000000000510654\tfrac{8}{10}$,
Perpendicular . . . $.000255663464646\tfrac{84}{10}$,
Hypothenuse . . . $.000255663464647\tfrac{359}{1000}$.

Polygon of 3145728 Sides.

Base . . . . . $.000127831732323\tfrac{679}{1000}$,
Perpendicular . . . $63.999999999872336\tfrac{1}{10}$,
Hypothenuse . . . $64.$

Base . . . . . $.000000000127663\tfrac{7}{10}$,
Perpendicular . . . $.000127831732323\tfrac{679}{1000}$,
Hypothenuse . . . $.000127831732323\tfrac{1155}{10000}$.

Now, the line drawn from the centre to an angle of the polygon, the perpendicular let fall on one of the equal sides, and half this side, form a right-angled triangle, in which there are known the base, which is half the side of the polygon, and the angle at the vertex, hence the perpendicular can be obtained.

The apothegm or perpendicular of the polygon of 3145728 sides being $63.999999999969646\tfrac{6}{10}$, and the perimeter $402.$ $123859659302850\tfrac{9}{10}$.

Then the half of the perpendicular, viz., 31.9999999999-84823$\frac{3}{10}$ × 402.123859659302850$\frac{9}{10}$ = 12867.963509091588-315, area of polygon.

Now as 63.999999999969646$\frac{6}{10}$, the apothegm or perpendicular of polygon 3145728 sides, is to its perimeter so is 64 the limit to the circumference of the circle, viz., 63.99999-9999969646$\frac{6}{10}$: 402.123859659302850$\frac{9}{10}$ :: 64 : 402.1238596-59493567.

Therefore 402.123859659493567 is the circumference of a circle 128 feet in diameter.

The diameter of a circle is to its circumference as the square of the radius to the area of the circle.

128 : 402.123859659493567 :: 4096 : 12867.963509103794144.

Therefore 12867.963509103794144 is the area of a circle 128 feet in diameter.

And the ratio is 1 to 3.141592653589793$\frac{63}{128}$.

The diameter of a circle is to the circumference in the ratio of 1 to 3.141592653589793$\frac{63}{128}$.

Therefore as 402.123859659493567 is to 128 so is the circumference to the diameter.

To Find the Diameter of a Circle.—Divide the circumference by 3.141592653589793$\frac{63}{128}$, and the quotient will be the diameter. For a short and near calculation, multiply the circumference by 113, and divide by 355.

To Find the Circumference of a Circle.—As 128 is to 402.12-3859659493567 so is the diameter to the circumference, that is, multiply the diameter by 402.123859659493567, and divide by 128, the product will be the circumference; and to shorten the calculation, multiply by 355, and divide by 113, which gives the answer nearly.

4

To Find the Area of a Circle.

RULE 1.—Multiply half the circumference by half the diameter, and the product will be the area.

RULE 2.—Multiply the square of the diameter by the decimal .785398163397448$\frac{6112}{16384}$, and the product will be the area; and to shorten the calculation, multiply by .7854, which will give the area nearly.

RULE 3.—Multiply the square of the circumference by the decimal .0795774715459476614 +, and the product will be the area; and to shorten the calculation, multiply by .07958, which will give the area nearly.

RULE 4.—Multiply the circumference by the radius, and divide the product by 2.

Area of the Circle by Per Cent.

The area of a circle can be determined by weight or centage. Cut from sheet-metal a square and circle, the diameter of the circle equal to the side of the square; weigh them, and take the difference. On trying the experiment, the result was as follows:

$$\text{Weight of square,} \ldots \ 238 \text{ grains.}$$
$$\text{Weight of circle,} \ldots \ 187 \text{ grains.}$$

The circle being $21\frac{43}{100}$ per cent. less than the square, therefore, by squaring the diameter of a circle, and subtracting $21\frac{46}{100}$ per cent., will give the area of the circle; for example:

$$64 \times 64 = 4096$$
$$\text{Less } 21\frac{46}{100} \text{ per cent.} \quad \underline{879}$$
$$3217$$

THE SQUARE OF THE CIRCLE.

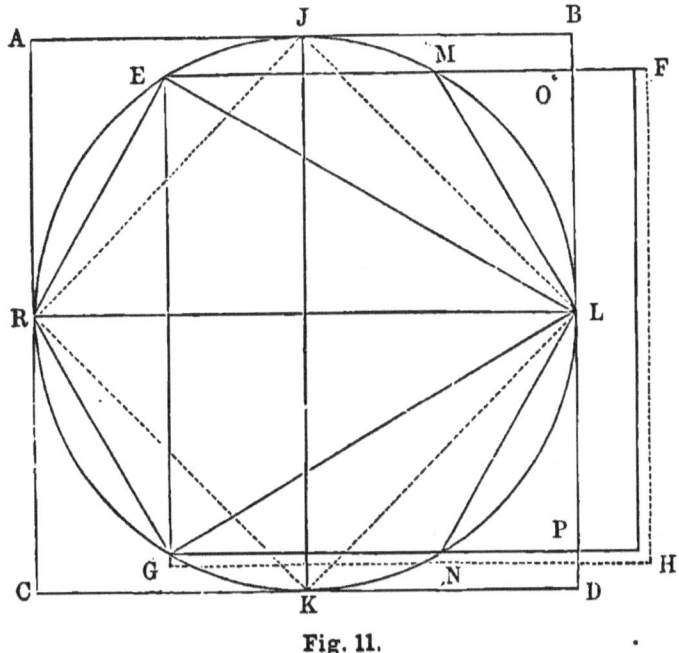

Fig. 11.

The circle, 64 feet diameter, circumference 201.0619298-29746783$\frac{64}{128}$, area 3216.990877275948536, equals a rectangle 32 × 100.530964914873391$\frac{96}{128}$.

The square, A B C D, sixty-four feet on a side, equals in area 4096 feet.

The square, R J L K, 45$\frac{218+}{1000}$ × 45$\frac{218+}{1000}$, or 45$\frac{23}{90}$ × 45$\frac{23}{90}$, that is, 45 × 45, and $\frac{23}{90}$ of 45 on the two angular sides, equals in area 2048 feet.

45.218	$45.\frac{23}{90}$	45
45.218	$45.\frac{23}{90}$	23
361 744	225	135
452 18	180	90
9043 6	11.45	90)1035(11
226090	11.45	90
180872	2048	135
2044667 524		90
		45

The square, E F G H, 55.424+ × 55.424+, or $55\frac{47}{110}$ × $55\frac{47}{110}$, in area 3072 feet, being three-fourths of the area of A B C D.

The square, E O G P, 48 × 55.424 +, or 48 × $55\frac{47}{110}$, is equal to the area of the hexagon 2660. $\frac{509\frac{10}{110}}{1000}$ feet.

The square, E F G H, including the dotted lines, is the equivalent of the circle, and the same area being 56.7185-$232\frac{32870103229}{113437801101}$ × $56.7185232\frac{32870103229}{113437801101}$ = in area 3216-.990877275948536 = a rectangle of 32 × 100.530964914873-$391\frac{96}{128}$.

The triangle, E G L, 55.424 + base, and 48 feet high, equals in area $1330\frac{352}{1000}$.

The hexagon, R E M L N G; area, $2660\frac{509\frac{10}{110}}{1000}$.

The square of the diameter of any circle is to its area as the perimeter of a square, described on the diameter of the circle, to its circumference, or the converse of the proposition. The area of any circle is to the square of its diameter as the circumference of the circle to the perimeter of a square described on its diameter. These facts may be demonstrated on any hypothetical data taken to represent the value of the circle; as, $4096 : 3216.9908772759\dot{+}8536 :: 256 : 201.0619298\text{-}29746783\frac{6\dot{4}}{128}$.

_____

THE SQUARE OF THE CIRCLE.

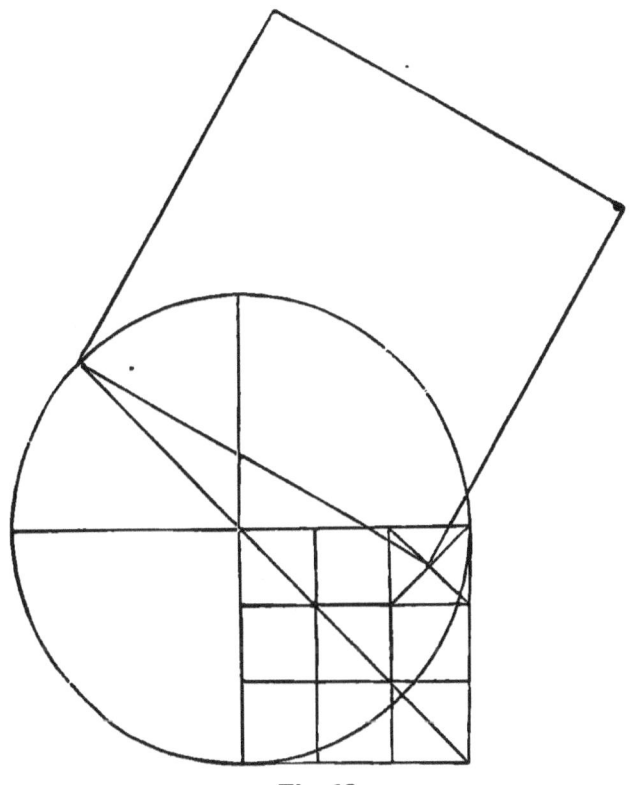

Fig. 12.

Let the diameter of this circle (Fig. 12) be 8 inches.

The circumference is $25.132741228718347\frac{120}{128}$.

The area of both is $50.265482457436695\frac{112}{124}$.

This circle and square are equal. The square is 7.089815403, and $\frac{8820643466}{141779630806}$ of 7.089815403 on two angular sides of the square.

4 *

128 : 402.123859659493567 :: 8 : 25.132741228718347⅛⅞⅞.

50.26.54.82.45.74.36.69.5⅕¼¼ (7.089815403

```
        49
1408    12654
   8    11264
14169     139082
    9     127521
141788    1156145
     8    1134304
1417961     2184174
      1     1417961
14179625    76621336
       5    70898125
141796304    572321169
        4    567185216
14179630803    51350535875
          3    42538892409
14179630806    8820643466
```

```
                          7089815403
                          8820643466
                         42538892418
                         42538892418
                         28359261612
                         21269446209
                         28359261612
                         42538892418
                         14179630806
                         56718523224
                         56718523224
```

```
14179630806 ) 62536733909618106798 ( 4410321733
              56718523224
               58182106856
               50718523224
                14635836321
                14179630806
                  45620551581
                  42538892418
                   30816501630
                   28359261612
                    24573300186
                    14179630806
                    103936693807
                     99257415642
                      40792781659
                      42538892418
                       42538892418
                       42538892418
```

```
        7089815403
        7089815403
       21269446209
       28359261612
       35449077015
        7089815403
       56918523224
       83808338627
       56718523224
       49628707821
5026548244 8616052409
         4410321733
         4410321733
5026548245743 6095875
```

Measurement by Parallel Lines.

To measure any figure, as a long irregular, curved, rectangle, square, or circular, by means of parallel lines drawn perpendicular, or at right angle to its greatest length; the greater the number of lines and spaces, the more perfect the measure. Divide the aggregate length of the lines by the number of spaces for the mean, and multiply that by the greatest length of the figure, and the answer will be the area.

EXAMPLE FIRST. — Let A B C D, Fig. 13, be a square, 32 feet; draw 21 lines perpendicular, as marked, extending them across the square, equal distances apart; the aggregate length of lines is 497¾ feet; divided by the number of spaces $22 = 22\frac{13}{22}$; this multiplied by the greatest width, viz., 45.254 +, equals 1023.872, the area of the square; the exact area is 1024.

EXAMPLE SECOND.—Draw 21 lines perpendicular to D E B, Fig. 13, triangle, D E B C; the aggregate length of the lines is 249 feet; divided by 22 spaces = $11\frac{7}{22}$; multiplied by the length, D E B, 45. 254 + = 512.193; the correct or exact area is 512 feet.

EXAMPLE THIRD. — The chord of one-fourth of a circle, the radius 32 feet, that is, D E B F, Fig. 13, divided in the same

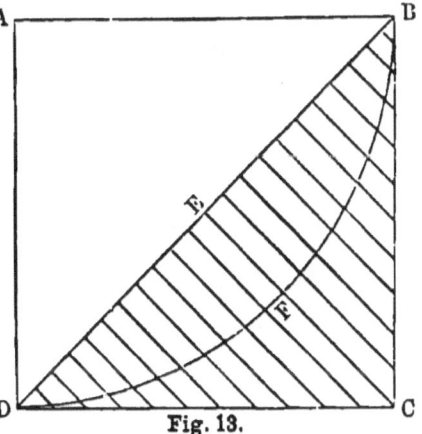

Fig. 13.

manner, and the lines aggregating $142\frac{8}{10}$ feet, divided by 22 spaces $= 6.\dfrac{10\frac{8}{10}}{22}$; multiplied by $45.254 + = 293.74$ feet area.

The correct or exact area is 292.247719318987134.

EXAMPLE FOURTH. — The corner triangle with curved side, D F B, divided in the same manner, and the lines aggregating $106\frac{6}{10}$ feet, divided by 22 spaces, equals $4.8\frac{12}{22}$; and multiplied by $45.254 + = 219.222$ feet area.

The correct and exact area is 219.752280681012866.

EXAMPLE FIFTH. — A circle of 32 feet diameter, divided into 39 lines and 40 spaces, the lines aggregating $1005\frac{6}{10}$ feet, divided by 40 spaces, equals $25\frac{14}{100}$, and multiplied by $32 = 804.48$.

The correct and exact area is 804.247719318987134.

EXAMPLE SIXTH. — Let Fig. 14, A B C D, represent a

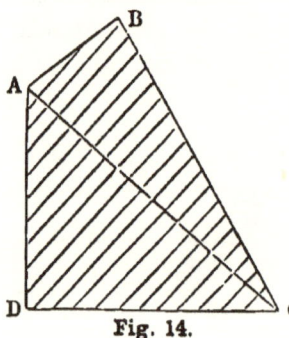

Fig. 14.

sail of the dimensions, viz., A B $11\frac{1}{2}$ feet, B C 31 feet, D C $22\frac{1}{2}$ feet, A D 22 feet. The largest line drawn across this will be from A to C, $31\frac{1}{2}$ feet; divided into 20 lines and 21 spaces, and the lines drawn at right angles to A C, aggregate $283\frac{5}{10}$ feet in length.

$283\frac{5}{10} \div 21 = 13\dfrac{10\frac{5}{10}}{24} = 423\frac{28}{100}$ square feet the area.

EXAMPLE SEVENTH. — Let a hexagon be 32 feet on a side; the largest length, or greatest diameter, will be 64 feet; draw 63 lines perpendicular to this. The lines will aggregate 1330 feet; divide by 64 spaces; the quotient multiplied by 64 gives the same answer, 1330 feet area.

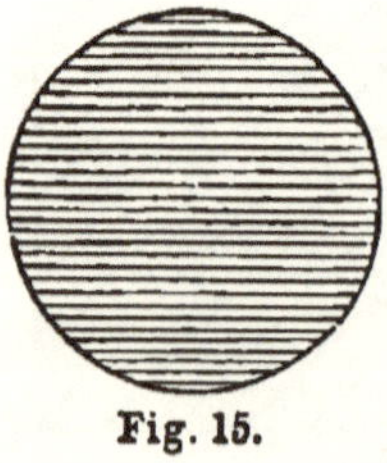

Let the circle, Fig. 15, be 16 feet in diameter; draw 31 lines; the aggregate length

Fig. 15.

of the lines is 402 feet; divided by 32 spaces $= 12\frac{18}{32}$; multiplied by 16 $= 201$ feet area.

Let Fig. 16, a ring or section of a circle, be 32 feet diameter, with a blank space in the centre 16 feet in diameter; draw 63 lines; the aggregate length of the lines is 1206 feet; divided by 64 spaces equals $188\frac{43}{64}$ feet; multiplied by 32 equals 603 feet area.

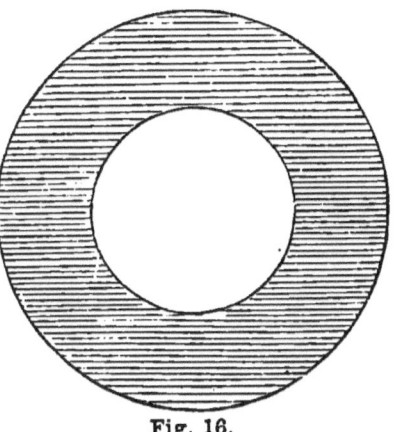

The square 32×64 lines (counting $\frac{1}{2}$ of each side one) $= 2048$ feet of lines, and the area is 1024 feet.

Fig. 16.

	Area.
Figure 15	201 feet.
Figure 16	603 "
Between Fig. 16 and the square .	220.6 "
	1024.6 "

In this manner the last two examples can be used as a proof or test of the exact contents of the circle, because doubling the diameter quadruples the area, and with the best instruments for ruling parallel lines an exact fractional measure can be obtained; if with imperfect ones, the above results are nearly exact. The calculations are made on a scale of 8 inches.

A quarter-sector of any circle is equal in area to a circle of half its diameter.

To Find the Area of a Triangle when the Three Sides are given.—First. Add the three sides together, and take half their sum.

Second. From this half sum subtract each side separately.

Third. Multiply together the half sum and each of the three remainders, and the product will be the square of the area of the triangle. Then extract the square root of this product for the required area. For example, take a triangle whose sides are 18, 24, 30.

18	36	36	36	36
24	18	24	30	18
30	18	12	6	288
2) 72				36
36				648
				12
				7776
				6
				46656

The square root of 46656 = 216 feet area.

The side or perpendicular of the equilateral triangle given to find the other dimension and area.

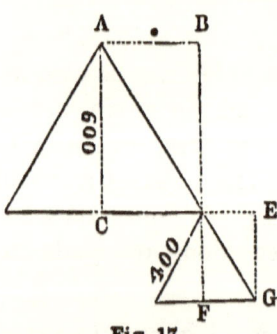

Fig. 17.

Let the large equilateral triangle be three times the size of the small one; then the perpendicular of the small triangle will be half the side of the large, and the perpendicular of the large triangle will be one and a half times the side of the small one.

The side of the small triangle is 400 feet.

The perpendicular of the large triangle is 600 feet.

We cannot give the exact dimensions of the side of the

large triangle, or the exact perpendicular of the small triangle; but we can give the exact square of the dimension of both, as follows:

The perpendic. of the large triangle, 600 × 600 = 360000.
"   "   "   " small   "   one-third = 120000.
The square root of 120000 = 346⁴⁸⁺₁₀₀.
The side of the small triangle, 400 × 400 = 160000.
"   "   "   " large   "   three times = 480000.
The square root of 480000 = 692.82+.

Side of square, D E, 200 × 200 = 40000.
Side of large square, A B, three times = 120000.
Side of large square, D B, 600 × 600 = 360000.
Side of small square, E G, one-third = 120000.

Therefore, for area of large triangle, A B, 120000 × B D 360000 = 43200000000, the square root of which equals 207846.096+, the area, and for small triangle area equals ⅓ 69282.032+.

RULE.—For the perpendicular of an equilateral triangle, multiply the side by the decimal .866025.

RULE.—For the side of an equilateral triangle, multiply the perpendicular by 1.1547.

To find the distance from A to B, when B only is accessible. (Fig. 18.)

Lay off a square with B D in range with the object, and stake each corner, say 12 feet square; extend the line D E to F in range with stake C and the object, say, for example, 4 feet, then as E F : E C :: B C : B A, or 4 : 12 :: 12 : 36.

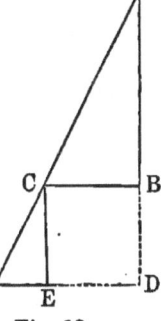

Fig. 18.

Table of Polygons.

NAMES.	Sides.	Angle at Vertex.	Angle at Centre.	Areas.
Trigon or Equilateral Triangle................	3	60°	120°	0.4330127
Tetragon or Square...	4	90°	90°	1.0000000
Pentagon	5	108°	72°	1.7204774
Hexagon..................	6	120°	60°	2.5980762
Heptagon ,...............	7	128° 34′ 29″	51° 25′ 71″	3.6339124
Octagon....................	8	135°	45°	4.8284271
Nonagon...................	9	140°	40°	6.1818242
Decagon	10	144°	36°	7.6942088
Undecagon...............	11	147° 16′ 36″	32° 43′ 64″	9.3656379
Dodecagon...............	12	150°	30°	11.1961524

To find the area of any regular polygon, multiply together one of its sides — that is, square one of its sides and multiply the product by the area, as above, and the quotient will be the area, unless the sides are decimals, when it will not answer.

To find the area of any parallelogram, multiply any side by the perpendicular height.

The diagonal of any parallelogram divides it into two equal triangles, and four diagonals divide it into four triangles of equal areas.

The four angles amount to four right angles, or 360°. Any two diagonally opposite angles are equal to each other, and having one angle, the other three can readily be found.

To find the area of any figure inclosed in four straight lines, draw two diagonals from the angles intersecting each other, multiply together the two diagonals and the natural sine of the least angle formed by the intersection, and divide the product by 2.

Or, draw a diagonal from opposite angles, and two perpendiculars from the vertex of opposite angles falling upon the diagonals; multiply the diagonals by the sum of the two perpendiculars, and half the product will be the area.

To find the diameter of a circle of the same area as a given ellipse, multiply together the two diameters, and take the square root of product $12 \times 20 = 240 = 15\frac{1}{2}$.

To Find the Area of an Ellipse.

Multiply the two diameters together; multiply the product by .7854.

EXAMPLE.— $12 \times 20 \times 7854 = 188.4960$.

The area of an ellipse is a mean proportional between the area of two circles described on its two diameters, therefore it may be found by multiplying together the areas of those two circles, and taking the square root.

To Find the Area of a Long, Irregular Figure.

Divide the space into equal distances; the more spaces the more correct. Measure these lines, add them together, and divide the sum by the number of lines for the mean breadth; then multiply that by the length of the figure, and the product will be the area.

Cylinders.

To find the surface, multiply the circumference by the perpendicular, to which add that of the ends.

To find the solidity, multiply the area of one end by the perpendicular; the solidity of a cylinder is three times that of a cone of the same base and height.

To find the solidity of any pyramid or cone, whether regular or irregular, right or oblique, multiply the area of its base by one-third of its perpendicular height. Every pyramid or cone has one-third of the solidity of either a prism or a cylinder, having the same area or base and the same

5   D

height, and one-half that of a hemisphere of the same base and height; a cone, hemisphere, and cylinder of the same base and height have solidities, as 1, 2, 3.

To find the surface of a right pyramid or right cone, multiply the outlines of its base or circumference by the slant height; take half the product. This will give the surface of the sides, and, if required, that of the base; the slant height must be measured from the apex to the middle of the base.

To find the solidity of any prismoid, add together the areas of the two parallel surfaces and four times the area of the section taken half-way between them, and parallel to them. Multiply the sum by the perpendicular distance between the two parallel sides, and divide the product by 6.

To find the solidity of a sphere, cube its diameter; multiply said cube by .5236, or cube its circumference, and multiply by .01689, or multiply its surface by its diameter, and divide by 6; the solidity of a sphere is two-thirds that of its circumscribing cylinder, or .5336 of that of its circumscribing cube.

To find the surface of a sphere, square its diameter; multiply said square by. 3183. The surface of a sphere is equal to four times the area of its great circle, or to the area of a circle whose diameter is twice as great as that of the sphere.

For the area of a square equal to an equilateral triangle, square the side, and multiply the quotient by the decimal .4330.

For the side of a square equal in area to a triangle, multiply the side of triangle by .658.

For the side of a triangle, multiply the side of the square by .15197.

To Find the Length of an Arc of a Circle containing any Number of Degrees.

Multiply the number of degrees in the given arc by $0.008726646259971\frac{2887}{4080}$, and the product by the diameter of the circle.

For since the circumference of a circle, whose diameter is 1, is $3.141592653589793\frac{63}{128}$, it follows that if 3.141592-$653589793\frac{63}{128}$ be divided by 360, the quotient will be the length of an arc of 1 degree, that is, $\dfrac{3.141592653589793\frac{63}{128}}{360}$

$= 0.008726646259971\frac{2887}{4080} =$ arc of one degree to the diameter 1.

If this be multiplied by the number of degrees in a given arc, the product will be the length of that arc in the circle, whose diameter is 1. If this product be then multiplied by the diameter of any circle, the product will be the length of the arc in that circle.

To find the length of an arc of 1 degree, the diameter being 128 feet.   *Ans.* $1.117010721276371\frac{7}{300}$.

To find the length of an arc of 1 degree, the diameter being the earth's diameter, $41846930.638\frac{10}{12}$ feet.

      Ans. $365183.360\frac{270}{360}$ feet.

For the side of a square equivalent to a circle, multiply the diameter by $.0886226\frac{1}{4}$ nearly.

The Earth's Equatorial Radius.
251081583.833 inches.
$20923465.319\frac{5}{12}$ feet.

The Earth's Equatorial Diameter.
502163167.666 inches.
$41846930.638\frac{10}{12}$ feet.

7925.555 statute miles.

6875⅟₃₀₇₁₈₃₃₄ geographical or nautical miles.

114⅞₄₀₂₁₂₃₈₅₉₀₅₅₄₉₃₅₆₇ degrees.

114 degrees, 35 minutes, 29 seconds, $\frac{6}{10}+$.

The Earth's Circumference at the Equator.

1577592118.442$\frac{8}{9}$ inches.

131466009.870 feet.

360 degrees.

24898.865½ statute miles.

21600 geographical or nautical miles.

69.$\frac{163}{1000}$ statute miles to 1 degree.

Area of the circumference equals a square of 7023 statute miles.

Area of the surface equals a square of 14048 statute miles.

Calculation of the Diameter and Circumference of the Earth in French Metres.

128 : 402.123859659493567 :: 12754646 : 40069120.

12754646 metres in diameter.

40069120 metres in circumference.

The one-fourth of the above, supposing it to be even forty millions, or the quadrant and the distance from the equator to the north pole, is adopted by the French Government as the fundamental standard for the metrical system of weights and measures.

The quadrant is divided by them into 10000000 equal parts, one of which is taken for one metre; and this unit is the foundation for all their weights and measures.

The Earth.

The earth is a planet in all respects similar to Venus and Mars, its nearest neighbors, which planets probably are inhabited the same as the earth. The form of the earth is not that of a sphere, but of an oblate spheroid, like possibly all of the planets. The eccentricity of the earth's orbit amounts to 0.1677110, but is subject to a small diminution, about 0.000041 in a period of 100 years, and this continued, must eventually result in a change of the form of the earth's orbit from an ellipse to a circle, unless some unknown external cause reacts after a period and changes or restores it again to its original condition.

The earth's heat increases at the rate say 100° a mile, many miles into the interior. At a distance of 12 miles into the interior, we have a red-hot heat; at a distance of 100 miles, the temperature will be so great as to melt the materials composing the crust of the earth. In fact, we have every reason to suppose that the crust has been cooling gradually for hundreds of thousands of years, and that our earth is nothing but the scoriæ from a fused mass from velocity, attraction, condensation, and gravitation.

We know that the earth is nearly a sphere by the appearance of a ship at sea; first the topmast looms up in the distance, then next the lower rigging, then the hull, when finally the whole and every part of the clipper-built vessel appears in all her beauty, walking the waters like a thing of life; also by the varying appearances of the constellations as we proceed northward or southward, and other indications. After the sun has set below the horizon and disappeared from the view of the inhabitants at the base of a high elevation, it can again be observed by ascending to the summit.

The earth turns on its axis in 23h. 56m. 4.09s., conse-

5 *

quently, we are travelling, by this diurnal motion, at the rate of 1040 miles per hour on our whirling vehicle the earth, in the direction from west to east in a circle, while at the same time we are travelling in an elliptical direction at the mean rate of 66.168 miles per hour,—a double motion somewhat similar to that of a top turning round, at the same time passing over the ground in a curved line.

The following Table of the Lengths of Days in Different Latitudes is from Mädler.

°	′	Hours.	°	′	Hours.
0	0	12	65	48	22
16	44	13	66	21	23
30	48	14	66	32	24
41	24	15	67	23	1 mo.
49	2	16	69	51	2 "
54	31	17	73	40	3 "
58	27	18	78	11	4 "
61	19	19	84	5	5 "
63	23	20	90	0	6 "
64.	50	21			

The 8646 hours which make up a year are according to the same authority.

AT THE EQUATOR.	AT THE POLES.
4348 hours day.	4389 hours day.
852 " twilight.	2370 " twilight.
3446 " night.	1887 " night.

Table Showing Difference of Time at 12 o'clock (noon) at New York.

New York . .	12.00 Noon.	Boston . . .	12.12 P. M.
Buffalo . . .	11.40 A. M.	Quebec . . .	12.12 "
Cincinnati . .	11.18 "	Portland . .	12.15 "
Chicago . . .	11.07 "	London . . .	4.55 "
St. Louis . .	10.55 "	Paris	5.05 "
San Francisco .	8.45 "	Rome . . .	5.45 "
New Orleans .	10.56 "	Constantinople	6.41 "
Washington .	1ł.48 "	Vienna . . .	6.00 "
Charleston . .	11.36 "	St. Petersburg	6.57 "
Havana . . .	11.25 "	Pekin, night .	12.40 A. M.

Time Table. — 60 seconds = 1 minute; 60 minutes = 1 hour; 24 hours = 1 day.

Measure or Division of the Circle.

60" seconds	=	1' minute,
60' minutes	=	1° degree,
30° degrees	=	1ˢ sign,
90° degrees	=	1 quadrant,
12 signs, or 360° degrees,	=	1 circumference.

The circumference of the globe, like every other circumference, is divided into 360 equal parts, called degrees; each degree is divided into 60 equal parts, called miles or minutes. Three miles are called a league.

Geographical or Nautical Measure.

6 feet	=	1 fathom,
110 fathoms, or 660 feet,	=	1 furlong,
120 fathoms	=	1 cable's length,
6086.389⅔ feet	=	1 nautical mile,

3 nautical miles   =   1 league,
20 leagues, or 60 geographical miles,   =   1 degree,
360 degrees the earth's circumference.

Gunter's Chain.

7.92 inches   =   1 link,
100 links   =   4 rods, or 22 yards.

Long Measure.

		in.	ft.	yd.	rds.	fur.
12 inches	= 1 foot,					
3 feet	= 1 yard,	36 =	3			
5½ yards	= 1 rod, pole, or perch,	198 =	16½ =	5½		
40 rods	= 1 furlong,	7920 =	660 =	220 =	40	
8 furlongs	= 1 statute mile,	63360 =	5280 =	1760 =	320 =	8
3 miles	= 1 league.					

Square Measure.

144 inches   =   1 square foot,
  9 feet   =   1   "   yard,
30¼ yards   =   1   "   rod, pole, or perch,
 40 rods   =   1   "   rood,

  4 roods   } =   1   "   acre,
10 square chains }

640 acres   =   1   "   mile.
43.560 square feet are one acre.
    16 square perches =   1   "   chain.
   5 yards wide by 968 long contain one acre.
10 "   "   "   484   "   "   "   "
20 "   "   "   242   "   "   "   "
40 "   "   "   121   "   "   "   "
  60 feet wide by 726 long contain one acre.
110 "   "   "   396   "   "   "   "
220 "   "   "   198   "   "   "   "

A strip of land one rod wide and one mile long is two acres.

Measure $2504\frac{1312}{2504}$, or $\frac{1311\frac{86}{100}}{2504}$ feet square, and you have one square acre. This last complex fraction is so nearly correct, that it makes a difference of only $\frac{5}{100}$ of one square inch in one hundred acres.

A standard English mile is 5.280 feet in length, 1760 yards, or 320 rods.

A U. S. Government township comprises 36 sections, each a mile square.

A section = 640 acres.

A quarter section, half a mile square = 160 acres.

An eighth section, half a mile long, north and south, and a quarter of a mile wide = 80 acres.

A sixteenth section, a quarter of a mile square = 40 acres.

Among the ancients, Aristarchus, of Samos, and Philolaus, maintained that not only did our globe rotate on its own axis, but that it revolved around the sun in twelve months. Other astronomers taught the same doctrine. The Egyptians taught the revolution of Mercury around the sun, and others gave the same motion to Mars, Jupiter, and Saturn.

' The origin of the division of the Zodiac into constellations is lost in obscurity; though attributed to the Greeks, it has been ascertained to be of greater antiquity, possibly the Hindoos or Chinese.

When on a railroad train making good time, say at the rate of forty miles an hour, fixed objects in view, such as houses, trees, fences, rocks, etc., are apparently hurrying past, while in reality we are hurrying past them; in like manner sun, moon, planets, and stars seem passing from east and west — this being in reality an apparent motion, because we are unconscious of our real motion of rapidly moving past them at the speed of over a thousand miles per hour. In this way we know that the earth revolves around the sun thousands of miles per hour in

D.	H.	M.	S.	
365	6	9	9.	Sidereal year,
365	5	48	46.	Tropical or solar year,
365	6	13	49.3	Anomalistic year.

In our reckoning, we make every fourth year to contain 366 days, and call it leap-year. Still greater accuracy requires, however, that the leap-day be dispensed with three times in every 400 years.

Whenever the number which denotes the year can be measured by 4, the year is leap-year, the centennial years excepted; and the centennial years divisible by 400 are also leap-years. The next centennial leap-year is 2000.

Calculation of the Distance of the Sun from the Earth.

When the earth is in perihelion, or nearest the sun, it then travels with its greatest velocity, and passes over an arc of 1° 01′ 9″.9 in a mean solar day.

When the earth is in aphelion, or farthest from the sun, it passes over an arc in the same time of only 57′ 11″.5.

The earth in its circuit or revolution around the sun sweeps over a distance of $580043925\frac{23}{100}$ miles, at an average speed or velocity of 18.38 miles per second, and mean distance of 66168 miles per hour, time and distance being the same and substituting a circle for the elliptic orbit.

Hence the equation

$$402.123859659493567 : 580043925.02 :: 128 : 184633716.$$ The last term being the diameter, half of this, viz., 92316858, is

the radius of the orbit and the distance of the sun from
the earth. This distance is from centre to centre, that is, distance plus the radius of the sun and earth. Now, deducting
half the diameter, or the radius of the sun and earth, 92316-
858 — 432374 = 91.884484, we have the true and exact distance of the earth from the sun, and thus this intricate problem yields itself up to demonstration.

$$91.884484 \quad = \quad \text{Mean} \quad \text{distance,}$$
$$93.425488 \quad = \quad \text{Maximum} \quad \text{``}$$
$$90.343480 \quad = \quad \text{Minimum} \quad \text{``}$$

The sun is a sphere, and is surrounded by an extensive
and rare atmosphere. It is self-luminous, emitting light and
heat, which are transmitted to a known distance of 2700
millions of miles.

The interval of time which intervenes from the moment
when the sun leaves a fixed star until it returns to it again
constitutes the sidereal year, and consists in solar time of
365 d. 6 h. 9 m. 9 s.; therefore the sidereal year is longer than
the mean solar year. The latter comprises the time between
two successive passages of the sun through the same equinox.
If the equinoxes were fixed points, then the period would be
the same as the sidereal revolution ; but since these points are
possessed of a retrograde motion from east to west of 50' 2"
annually, and the sun returns to the equinox sooner every
year by a period of time of 20 m. 23 s., the mean solar year
is therefore 20 m. 23 s. shorter than the sidereal year, in consequence of the motions of the equinoctial points not being
uniform.

According to M. Delambre, if the earth be supposed to
start from perihelion, it will require a longer interval of time
than the sidereal period to reach perihelion again ; and the
excess will be equal to the time necessary for the earth to
describe 11' 8" of its orbit; this it would do in 4 m. 39.7 s.,

which quantity must be added to the sidereal before we can ascertain the interval between two successive returns to perihelion.

The result, then, is a period of 365.259581 mean solar days, which is called the anomalistic year.

Multiply the earth's diameter (7925.555) by 108, and we have 855960, nearly the sun's diameter in miles.

Multiply the sun's diameter (856.822, this diameter of the sun is calculated to correspond to my distance) by 108, and we have 92.536776, nearly the mean distance of the earth from the sun.

Multiply the moon's diameter (2160) by 108, and we have 233280, nearly the mean distance of the moon from the earth.

The proportion of the sun's heat that reaches us is only $\frac{1}{2300000000}$. What the whole amount must be is beyond all human comprehension. Our annual share is sufficient to melt a layer of ice covering the surface of the earth to the depth of thirty-eight yards in thickness; and according to Pouillet, another calculation determines the direct light of the sun to be equal to that afforded by 5563 wax candles of moderate size, supposed to be placed at a distance of one foot from the observer. The light of the moon being probably equal to that of only one candle at a distance of twelve feet, it follows that the light of the sun exceeds that of the moon 801072 times, according to Wollasten; Zollner's ratio is 618000.

The sun's mass, or attractive power, exceeds that of the earth 314.760 times.

The two points where the celestial equator intersects the elliptic are called the equinoxes (from *æquus*, equal, and *nox*, a night), because, when the sun is at these points, day and night are theoretically equal.

The points midway between these points are called the solstices (from *sol*, the sun, and *stare*, to stand still), because the sun, when it has reached these neutral points, has obtained its greatest declination north or south, as the case may be).

Kepler's Discoveries.

The square of Jupiter's period is to the square of Saturn's period as the cube of Jupiter's distance is to the cube of Saturn's distance.

Kepler's first Law.—That all orbits are ellipses, with the sun or primary body in one of the foci of the ellipse.

Second Law.— Equal areas of their orbits are described by the planets in equal times; the orbit through its entire course is so balanced that the rapidity is exactly proportional to the nearness, and the slowness by the distance, in reference to each other; so that equal areas of the orbit are described by the planets in equal times.

Third Law.— That the squares of the periodic times of the planets are proportional to the cubes of their major axis or of their mean distances. (Seventy-eight years after Kepler's discoveries, Newton discovered the law of gravitation.) If, for illustrating this Law, we take Mars and the Earth for an example, we find that the period of revolution of Mars is 686.23 days, and that of the Earth 365.6 days, and these periods, when squared, will give for the Earth $8766 \times 8766 = 76842756$ hours, and for Mars $16487 \times 16487 = 271821169$, and that the square of the period of Mars is therefore rather more than three times that of the earth.

In like manner, if we take the mean distance of Mars, which is 140031954 millions of miles, and that of the Earth, which is 91.884484 millions of miles, we have in cubing these mean distances $140 \times 140 \times 140 = 2744000$ for Mars', and $91\frac{8}{10} \times 91\frac{8}{10} \times 91\frac{8}{10} = 773620$ for the Earth's; and dividing these results by each other, we shall find that

6

the cube of the mean distance of Mars is rather more than three times the cube of the earth's mean distance.

This proportion between the squares of the times and the cubes of the mean distances is found so universally to prevail with regard to all the planets and satellites, and other celestial bodies moving in orbits, that it has all the appearance of an absolute law.

A Day.

A day is either natural or artificial. A natural day is the space of time which elapses while the sun goes from any meridian until he arrives at the same again, or it is the time contained from noon to noon, or the same hour again.

An artificial day is the time between the sun's rising and setting, to which is opposed the night, the time the sun is hid under the horizon.

All nations do not begin their day and reckon their hours alike. In most places in Europe the day is reckoned to begin at midnight, from whence is counted 12 hours till noon, then 12 more until next midnight, which make a day; yet astronomers commonly begin their day at noon, and reckon twenty-four hours until next noon.

The Babylonians began their day at sunrising, and reckoned twenty-four hours until he rose again. This we call the Babylonish hours.

The Jews and the Romans divided the artificial days and nights into twelve equal parts. These were termed the Jewish hours, and were of different lengths, according to the seasons of the year — a Jewish hour in summer being longer than one in winter, and a night hour shorter, and the hours were styled the first, second, etc., of the day or night, so that midday always fell on the sixth hour of the day. These hours were also called planetary hours, because in every hour one of the seven planets was supposed to preside over the world, and to take it by turns. The first hour after sunrising on Sunday was allotted to the Sun, the next to Venus, the third to Mercury, and the rest in order to the Moon, Saturn, Jupiter, and Mars. By this means, on the first hour of the next day the Moon presided, and so gave the name to that day; and so the seven days, by this method, had names given them from the planets that were supposed to govern on the first hour.

All nations that have any notion of religion set apart one day in seven for public worship. The day solemnized by Christians is Sunday, or the first day of the week, being that on which our Saviour rose from the grave, and on which the apostles afterwards used more particularly to assemble together to perform divine worship.

The Jews observed Saturday, or the seventh day of the week, for their Sabbath or day of rest, being that appointed in the fourth commandment under the law. The Turks perform their religious ceremonies on Friday.

A Month.

A month is that space of time measured by the moon in its course around the earth. A lunar month is periodical or synodical. A periodical is that space of time the moon takes to perform her course from one point of the elliptic till she arrives at the same again, which is 27 days and some odd hours; and a synodical month is the time between one new moon and the next, about 29½ days. But a civil month is different from these, and consists of a number, according to the laws and customs of the country wherein they are observed.

A Year.

The civil year is the same with the political established by the laws of a country, and is either movable or immovable; the movable year consists of 365 days, being less than the tropical, and is called the Egyptian year, because observed in that country.

The Romans divided the year into twelve calendar months, to which they gave particular names, that are still retained by most nations.

The year is also divided into four quarters or seasons, Spring, Summer, Autumn, and Winter. These quarters are properly made when the sun enters into the equinoctial and solstitial points of the elliptic; but in civil uses they are differently reckoned, according to the customs of the several countries. In the United States and England we commonly reckon the first day of January to be the first in the year, which is commonly called New-Year's Day. In ecclesiastical affairs the day is reckoned to commence on Lady-Day, which is the 25th March, and from thence to midsummer day, which is the 24th June, is reckoned the first quarter; from midsummer day to Michaelmas-Day, which is the 29th of September, is the second quarter; the third quarter is reckoned from Michaelmas-Day to Christmas-Day, which is the 25th of December, and from Christmas-Day to Lady-Day is reckoned the last quarter in the year. In common affairs, a quarter is reckoned from a certain day to the fourth month following; sometimes a month is reckoned four weeks or 28 days, and so a quarter 12 weeks. To all the inhabitants in the {Northern / Southern} hemispheres, their midsummer is properly when the sun is in the tropics of {Cancer / Capricorn} and their midwinter at the opposite time of the year; but those who live under the equinoctial have two winters, viz., when the sun is in either tropic; although properly there is no season that may be called winter in those parts of the world.

The Egyptian year of 365 days being less than the true solar year by almost six hours, it follows that four such years are less than four solar years by a whole day, and therefore in 365 times four years, that is in 1460 years, the beginning of the year moves through all seasons. To remedy this inconvenience, Julius Cæsar (considering that the six hours that remain at the end of every year will in four years make a natural day) ordered that every fourth year should have an intercalary day, which therefore consists of 366 days. The day added was put in the month of February by postponing St. Matthias's day, which in common years falls on the 24th, to the 25th of said month; all the fixed feasts in the year from thenceforward falling a week-day later than they would otherwise.

According to the Roman way of reckoning, the 24th of February was the sixth of the Calends of March; and it was ordered that for this year there should be two sixths, or that the sixth of the Calends of March should be twice repeated, upon which account the year was called Bissextile, which we now call the leap-year. To find whether the year of our Lord be leap-year, or the first, second, or third after, divide it by four, and the remainder, if any, shows how many years it is after leap-year; but if there be no remainder, then that year is leap-year.

This method of reckoning the year, making the common year to consist of 365 days, and every fourth year to have 366 days, is called the Julian account, or the old style.

But the time appointed by Julius Cæsar for the length of a solar year was too much. At the time of the Council of Nice (where the terms were settled for observing Easter), the vernal equinox fell upon the 21st of March; but by its falling backwards 11 minutes every year, it was found

6 * E

that in A. D. 1582, when the calendar was corrected, the sun entered the equinoctial circle on the 11th of March, having departed ten whole days from its former place in the year. Pope Gregory XIII., therefore, designing to place the equinoxes in their former situation with respect to the year, took these ten days out of the calendar, and ordered that the 11th of March should be reckoned as the 21st. To prevent the seasons of the year from going backwards for the future, he ordered that every hundredth year, which in the Julian form was to be a Bissextile, should be a common year, and consist only of 365 days; but that being too much, every fourth hundred was to remain Bissextile. This form of reckoning, being established by the authority of Pope Gregory XIII., is called the Gregorian account, or the new style, and is observed in all Christian countries.

In the year 1752, when the Gregorian calendar was adopted by the English government, eleven days had to be dropped out of the almanac. Even the Gregorian system does not give exactly correct results, but by a slight change can be so adjusted that it will not vary the commencement of the year more than a day in one hundred thousand years.

The Stars.

The fixed stars are those bright and shining bodies which on a clear night appear to us everywhere dispersed through the boundless regions of space. They are termed fixed because they have been found to keep the same immutable distance one from another in all ages, without having any of the motions observed in the planets. The fixed stars are all placed at such immense distances from us that the best of telescopes represent them no larger than points, without having any diameters.

It is evident that all the stars are luminous bodies, and shine with their own proper and native light, else they could not be seen at such a great distance. For the satellites of Jupiter and Saturn, although they appear under considerable angles through good telescopes, are altogether invisible to the naked eye. The distance between us and the sun is vastly large when compared to the diameter of the earth; yet it is nothing, when compared with the enormous distance of the fixed stars, for the whole diameter of the earth's annual orbit appears from the nearest fixed stars no larger than a point, and the fixed stars are at least 100000 times farther from us than we are from the sun.

Hence it follows that though we approach nearer to some fixed stars at one time of the year than we do at the opposite period, and that by the whole length of the diameter of the earth's orbit; yet this distance being so small in comparison with the distance of the fixed stars, their magnitudes or positions cannot thereby be sensibly altered; therefore we may always, without error, suppose ourselves to be in the same centre of the heavens, since we always have the same visible prospect of the stars without any alteration.

If a spectator were placed as near to any fixed star as we are to the sun, he would then observe a body as large, and every way like, as the sun appears to us, and our sun would appear to him no larger than a fixed star; and undoubtedly he would reckon the sun as one of them in numbering the stars; therefore, since the sun differs nothing from a fixed star, the fixed stars may be reckoned as so many suns.

It is not reasonable to suppose that all the fixed stars are placed at the same distance from us; but it is more probable that they are everywhere interspersed through the vast indefinite space of the universe, and that there may be as great a distance between any two of them as there is between our sun and the nearest fixed star.

Hence it follows why they appear to us of different magnitudes, not because they are at different distances from us, those that are nearest excelling in brightness and lustre, those that are more remote, which give a fainter light, and appear smaller to the eye.

Astronomers distribute the stars into several orders or classes, — those that are nearest to us, and appear brightest to the eye, are called stars of the first magnitude; those that are next in brightness and lustre are called stars of the second magnitude; those next the third, and so continued to stars of the sixth magnitude, which are the smallest that can be discovered by the naked eye. There are infinite numbers of smaller stars that can be seen through telescopes, but these are not reduced to any of the six orders, being called telescopical stars.

The ancient astronomers, that they might distinguish the stars in regard to their situation and position to each other, divided the whole starry firmament into several asterisms, or systems of stars, consisting of those that are near to one another. These asterisms are called constellations, and are fancifully compared with the forms of some animals, as men, lions, bears, serpents etc., or to the images of some known things, as of a crown, a harp, a triangle, etc.

The starry firmament was divided by the ancients into 48 images or constellations, twelve of which they placed in that part of the heavens wherein are the planes of the planetary orbits, which part is called the Zodiac, because most of the constellations placed therein were supposed to resemble some living creature. The two regions of the heavens that are on each side of the Zodiac are called the north and south parts of the heavens. The ancients placed these particular constellations or figures in the heavens either to commemorate the deeds of some great man or of some notable exploit or action, or else took them from the fables of their religion;

and the modern astronomers still retain them, to avoid the confusion that would arise by making new ones when they compared the modern observations with the old ones. Some of the principal stars have particular names given them, as Sirius, Arcturus, etc.

There are also several stars that are not reduced into constellations, and these are called unformed stars.

Besides the stars visible to the naked eye, there is a very remarkable space in the heavens called the galaxy, or milky way. This is a broad circle, of a whitish hue like milk, going quite round the heavens, and consisting of an infinite number of small stars, visible through a telescope, though not discernible to the naked eye by reason of their exceeding faintness; yet with their light they combine to illustrate that part of the heavens where they are, and to cause that shining whiteness.

The places of the fixed stars, or their relative situation one to another, have been carefully observed by astronomers and collected in catalogues.

The first among the Greeks who reduced the stars into a catalogue was Hipparcus, who, from his own observations and of those who lived before him, inserted 1022 stars upon his catalogue, about 120 years before the Christian era. This has been increased to about three thousand discernible by the naked eye.

It may seem strange to some persons that there are no more than this number of stars visible to the naked eye, for sometimes on a clear night they appear to be innumerable; but this is only a deception of our sight, arising from their constant sparkling, while we look upon them confusedly, without reducing them into any order, for there can seldom be seen above 1000 stars in the whole heavens, with the naked eye, at the same time; and if we distinctly view them, we will not find one but what is marked upon a good celestial globe.

Although the number of stars that can be discerned by the naked eye are so few, yet there are many more which are beyond the reach of our vision. Through telescopes they appear in vast multitudes dispersed throughout the firmament, and the better the glasses the more appear.

Those who think that all of these glorious bodies were created for no other purpose than to impart to us a little dim light must entertain a very weak idea of the Divine Wisdom. We receive more light from the moon itself, because of its comparative nearness, than from all the stars put together. Nevertheless, this comparison does not in the least detract from the actual resplendent character or reality, no, not one iota.

The firmament is a creation of the Almighty and his diadem or crown of glory, the sparkling stars the jewels thereof. Seeing that the planets are subject to the same laws of motion with our earth, and some of them are not only equal, but vastly exceed it in magnitude, it is not unreasonable to suppose that they are all inhabitable worlds. And since the fixed stars are no way behind our sun, either in size or lustre, is it not probable that some, if not all of them, have a system of planetary worlds turning around them as we do around our sun; and if we ascend as far as the smallest star we can see, shall we not then discern innumerably more of these glorious bodies which now are altogether invisible to us, and so *ad infinitum* through the boundless space of the universe?

What a magnificent idea must this raise in us of the Divine Being! who is everywhere and at all times present, displaying his power, wisdom, and goodness amongst all his creatures!

Length of a Degree of Longitude in Different Latitudes and at the Level of the Sea.

These lengths are in common land or statute miles of 5280 feet. Since the figure of the earth has never been precisely ascertained, these are but close approximations.

Degree of Lat.	Miles.	Degree of Lat.	Miles.	Degree of Lat.	Miles.	Degree of Lat.	Miles.
0	69.17	22	64.18	44	49.83	64	30.38
2	69.13	24	63.29	46	48.13	66	28.22
4	69.00	26	62.22	48	46.31	68	25.98
6	68.80	28	61.12	50	44.54	70	23.73
8	68.50	30	59.93	52	42.66	72	21.44
10	68.13	32	58.68	54	40.73	74	19.13
12	67.67	34	57.39	56	38.74	76	16.80
14	67.13	36	56.00	58	36.70	78	14.45
16	66.50	38	54.57	60	30.64	80	12.08
18	65.80	40	53.06	62	32.53	82	9.69
20	65.04	42	51.47				

Limits of Vegetation in the Temperate Zones.

The vine ceases to grow at about 2300 feet above the level of the sea; Indian corn, at 2800; oak, at 3350; walnut, at 3600; ash, at 4800; yellow pine, at 6200; and fir, at 6700.

Perpetual snow under the equator at 15800 feet above the level of the sea; in latitude 45°, at 8400; and in latitude 65°, at 5000.

Equation of Payments.

RULE 1.—Multiply each debt by the number of days, counting from the time of the earliest date to the date of each sum respectively; then divide the sum of these products by the sum of the account: the quotient will give the number of.days to count, from the first debt, for the average date of the account.

NOTE. — Since the first debt is the period from which the average time is computed, it must be left out of the dividend, but must be included in the divisor.

Rule for unequal credits, viz., for six, nine, and twelve months. — Find the date when due, and apply the rule as above, which will give the average date when due.

RULE 2. — Take the difference of time between each payment and the last, which arrange and multiply as in example below; then divide the sum of this product by the whole amount, and the quotient will be the number of days to count back from the last date in the arrangement.

			Sums.	Days.	Products.
First amount fell due	February 19,	900 × 185 =	166500
Second " "	March 1,	4,000 × 175 =	700000
Third " "	March 3,	3,500 × 173 =	605500
Fourth " "	March 6,	8,640 × 170 =	1468800
Fifth " "	March 18,	6,300 × 158 =	995400
Sixth " "	April 29,	5,000 × 116 =	580000
Seventh " "	August 23,	500 × 1 =	500

28,840   ) 4516700

Quotient, 157 days nearly.

So the $28,840 fall due on March 19th, counting back 157 days from August 23d.

Explanation of the table of days showing the number of days from any date in one month to the same date in any other month.

EXAMPLE. — How many days from September 9th to March 9th? Look for Sept. at the left hand, and March at the top — in the angle is 181. In leap-year, *add one day if February is* included.

Table of Days for Interest, etc.

FROM	To Jan.	To Feb.	To Mar.	To Apr.	To May.	To June	To July	To Aug.	To Sept.	To Oct.	To Nov.	To Dec.
January....	365	31	59	90	120	151	181	212	243	273	304	334
February..	334	365	28	59	89	120	150	181	212	242	273	303
March......	306	337	365	31	61	92	122	153	184	214	245	275
April.......	275	306	334	365	30	61	91	122	153	183	214	244
May.........	245	276	304	335	365	31	61	92	123	153	184	214
June........	214	245	273	304	334	365	30	61	92	122	153	183
July........	184	215	243	274	304	335	365	31	62	92	123	153
August.....	153	184	212	243	273	304	334	365	31	61	92	122
September	122	153	181	212	242	273	303	334	365	30	61	91
October....	92	123	151	182	212	243	273	304	335	365	31	61
November	61	92	120	151	181	212	242	273	304	334	365	30
December.	31	62	90	121	151	182	212	243	274	304	335	365

Table of Experiments made by the British Admiralty with Wire and Hempen Rope and Chain—Comparative Strength.

Circumference of Wire Rope.	Circumference of Hempen Rope.	Diameter of Chain.	Breaking Weight.
Inches.	Inches.	Inches.	Lbs.
2	5	$\frac{1}{2}$	14,224
3	8	$1\frac{1}{16}$	26,880
4	10	$\frac{3}{32}$	43,232

Explanation of the Following Interest Table.

The principal, beginning at 1 dollar and progressing to 2000, is to be found at the top of the page; the time in the left hand margin of each page begins at the top with 1 day, proceeding down in regular order to 30 days, and after the blank line the months begin and proceed in like manner, so that we have in view on the same page the interest of the sum required for years, months, and days; if the interest is required for two years, double the year.

EXAMPLE. — Required the interest for 2 years, 9 months, 17 days on 50 dollars. Turn to 50, at the top of column, page 79, and you have at the bottom,

in line with 12 mos., \$3.00 × 2 =   \$6.00
"   "   "   9 mos.   2.25
"   "   "   17 days   .14

These added together is the interest required,   \$8.39

If we want the interest on part of a dollar, turn to the cent table on page 86.

If 1 per cent. is required, divide 839 by six.

If 2 per cent. is required, divide by three.

If 4 per cent. is required, deduct one-third.

If 5 per cent. is required, deduct one-sixth.

If 7 per cent. is required, add one-sixth.

If 8 per cent. is required, add one-third to the above.

If 9 per cent. is required, add one-half.

If 10 per cent is required, add two-thirds.

If 11 per cent is required, add five-sixths.

If 12 per cent is required, multiply by two.

Interest Table at 6 Per Cent.

Days.	\[DOLLARS\] 1	2	3	4	5	6	7	8	9	10	11
1	0	0	0	0	0	0	0	0	0	0	0
2	0	0	0	0	0	0	0	0	0	0	0
3	0	0	0	0	0	0	0	0	0	0	0
4	0	0	0	0	0	0	0	0	1	1	1
5	0	0	0	0	0	0	0	1	1	1	1
6	0	0	0	0	0	1	1	1	1	1	1
7	0	0	0	0	0	1	1	1	1	1	1
8	0	0	0	0	1	1	1	1	1	1	1
9	0	0	0	0	1	1	1	1	1	1	2
10	0	0	0	1	1	1	1	1	1	2	2
11	0	0	1	1	1	1	1	2	2	2	2
12	0	0	1	1	1	1	1	2	2	2	2
13	0	0	1	1	1	1	2	2	2	2	2
14	0	0	1	1	1	1	2	2	2	2	2
15	0	0	1	1	1	1	2	2	2	2	2
16	0	0	1	1	1	1	2	2	2	3	3
17	0	0	1	1	1	2	2	2	3	3	3
18	0	0	1	1	1	2	2	2	3	3	3
19	0	0	1	1	1	2	2	3	3	3	3
20	0	0	1	1	1	2	2	3	3	3	4
21	0	0	1	1	2	2	2	3	3	4	4
22	0	1	1	1	2	2	3	3	3	4	4
23	0	1	1	1	2	2	3	3	3	4	4
24	0	1	1	2	2	2	3	3	3	4	4
25	0	1	1	2	2	2	3	3	4	4	5
26	0	1	1	2	2	3	3	3	4	4	5
27	0	1	1	2	2	3	3	4	4	5	5
28	0	1	1	2	2	3	3	4	4	5	5
29	0	1	1	2	2	3	3	4	4	5	5
30	0	1	1	2	2	3	3	4	4	5	5
Mos.											
1	0	1	1	2	2	3	3	4	4	5	5
2	1	2	3	4	5	6	7	8	9	10	11
3	1	3	4	6	7	9	10	12	13	15	16
4	2	4	6	8	10	12	14	16	18	20	22
5	2	5	7	10	12	15	17	20	22	25	27
6	3	6	9	12	15	18	21	24	27	30	33
7	3	7	10	14	17	21	24	28	31	35	38
8	4	8	12	16	20	24	28	32	36	40	44
9	4	9	13	18	22	27	31	36	40	45	49
10	5	10	15	20	25	30	35	40	45	50	55
11	5	11	16	22	27	33	38	44	49	55	60
12	6	12	18	24	30	36	42	48	54	60	66

Interest Table at 6 Per Cent.—*Continued.*

Days.	12	13	14	15	16	17	18	19	20	21	22
1	0	0	0	0	0	0	0	0	0	0	0
2	0	0	0	1	1	1	1	1	1	1	1
3	1	1	1	1	1	1	1	1	1	1	1
4	1	1	1	1	1	1	1	1	1	1	1
5	1	1	1	1	1	1	1	1	2	2	2
6	1	1	1	1	2	2	2	2	2	2	2
7	1	1	2	2	2	2	2	2	2	2	3
8	2	2	2	2	2	2	2	2	3	3	3
9	2	2	2	2	2	3	3	3	3	3	3
10	2	2	2	2	3	3	3	3	3	3	4
11	2	2	3	3	3	3	3	3	4	4	4
12	2	3	3	3	3	3	4	4	4	4	4
13	3	3	3	3	3	4	4	4	4	4	5
14	3	3	3	3	4	4	4	4	5	5	5
15	3	3	3	3	4	4	4	5	5	5	5
16	3	3	4	4	4	4	5	5	5	6	6
17	3	4	4	4	4	5	5	5	6	6	6
18	4	4	4	5	5	5	5	6	6	6	7
19	4	4	4	5	5	5	6	6	6	7	7
20	4	4	5	5	5	6	6	6	7	7	7
21	4	5	5	5	6	6	6	7	7	7	8
22	4	5	5	5	6	6	6	7	7	8	8
23	5	5	5	6	6	7	7	7	8	8	8
24	5	5	6	6	6	7	7	7	8	8	9
25	5	5	6	6	7	7	8	8	8	9	9
26	5	6	6	7	7	7	8	8	9	9	10
27	5	6	6	7	7	7	8	8	9	9	10
28	6	6	7	7	8	8	8	9	9	10	10
29	6	6	7	7	8	8	8	9	10	10	11
30	6	6	7	7	8	8	9	9	10	10	11
Mos.											
1	6	6	7	7	8	8	9	9	10	10	11
2	12	13	14	15	16	17	18	19	20	21	22
3	18	20	21	23	24	25	27	28	30	31	33
4	24	26	28	30	32	34	36	38	40	42	44
5	30	32	35	37	40	42	45	47	50	52	55
6	36	39	42	45	48	51	54	57	60	63	66
7	42	45	49	52	56	59	63	66	70	73	77
8	48	52	56	60	64	68	72	76	80	84	88
9	54	58	63	67	72	76	81	85	90	94	99
10	60	65	70	75	80	85	90	95	100	105	110
11	66	71	77	82	88	93	99	104	110	115	121
12	72	78	84	90	96	102	108	114	120	126	132

Interest Table at 6 Per Cent.—*Continued.*

Days.	23	24	25	26	27	28	29	30	31	32	33
1	0	0	0	0	0	0	0	0	1	1	1
2	1	1	1	1	1	1	1	1	1	1	1
3	1	1	1	1	1	1	1	1	2	2	2
4	2	2	2	2	2	2	2	2	2	2	2
5	2	2	2	2	2	2	2	2	3	3	3
6	2	2	2	3	3	3	3	3	3	3	3
7	3	3	3	3	3	3	3	3	4	4	4
8	3	3	3	3	4	4	4	4	4	4	4
9	3	4	4	4	4	4	4	4	5	5	5
10	4	4	4	4	4	5	5	5	5	5	5
11	4	4	5	5	5	5	5	5	6	6	6
12	5	5	5	5	5	6	6	6	6	6	7
13	5	5	5	6	6	6	6	6	7	7	7
14	5	6	6	6	6	6	7	7	7	7	8
15	6	6	6	6	7	7	7	7	8	8	8
16	6	6	7	7	7	7	8	8	8	8	9
17	6	7	7	7	8	8	8	8	9	9	9
18	7	7	7	8	8	8	9	9	9	9	10
19	7	7	8	8	8	9	9	9	10	10	10
20	8	8	8	9	9	9	10	10	10	11	11
21	8	8	9	9	9	10	10	10	11	11	12
22	8	9	9	10	10	10	11	11	11	11	12
23	9	9	10	10	10	11	11	11	12	12	13
24	9	10	10	10	11	11	12	12	12	13	13
25	10	10	10	11	11	12	12	12	13	13	14
26	10	10	11	11	12	12	13	13	13	14	14
27	10	11	11	12	12	13	13	14	14	14	15
28	11	11	12	12	13	13	14	14	15	15	15
29	11	12	12	13	13	14	14	11	15	15	16
30	11	12	12	13	13	14	14	15	15	16	16
Mos.											
1	11	12	12	13	13	14	14	15	15	16	16
2	23	24	25	26	27	28	29	30	31	32	33
3	34	36	37	39	41	42	44	45	47	48	50
4	46	48	50	52	54	56	58	60	62	64	66
5	57	60	62	65	68	70	73	75	78	80	83
6	69	72	75	78	81	84	87	90	93	96	99
7	80	84	87	91	95	98	102	105	109	112	116
8	92	96	100	104	108	112	116	120	124	128	132
9	103	108	112	117	122	126	131	135	140	144	149
10	115	120	125	130	135	140	145	150	155	160	165
11	126	132	137	143	149	154	160	165	171	176	182
12	138	144	150	156	162	168	174	180	186	192	198

7 *

Interest Table at 6 Per Cent.—*Continued.*

Days.	DOLLARS.										
	34	35	36	37	38	39	40	41	42	43	44
1	1	1	1	1	1	1	1	1	1	1	1
2	1	1	1	1	1	1	1	1	1	1	1
3	2	2	2	2	2	2	2	2	2	2	2
4	2	2	2	2	2	3	3	3	3	3	3
5	3	3	3	3	3	3	3	3	3	4	4
6	3	3	4	4	4	4	4	4	4	4	4
7	4	4	4	4	4	4	5	5	5	5	5
8	4	5	5	5	5	5	5	5	6	6	6
9	5	5	5	5	6	6	6	6	6	6	7
10	6	6	6	6	6	6	7	7	7	7	7
11	6	6	7	7	7	7	7	7	8	8	8
12	7	7	7	7	7	8	8	8	8	8	9
13	7	7	8	8	8	8	9	9	9	9	10
14	8	8	8	9	9	9	9	10	10	10	10
15	8	9	9	9	10	10	10	10	11	11	11
16	9	9	9	10	10	10	11	11	11	11	12
17	10	10	10	10	11	11	11	12	12	12	12
18	10	10	11	11	11	12	12	12	13	13	13
19	11	11	11	12	12	12	13	13	13	14	11
20	11	12	12	12	13	13	13	14	14	14	15
21	12	12	13	13	13	14	14	14	15	15	15
22	12	13	13	14	14	14	15	15	15	16	16
23	13	13	14	14	15	15	15	16	16	16	17
24	14	14	14	15	15	16	16	16	17	17	18
25	14	15	15	15	16	16	17	17	17	18	18
26	15	15	16	16	16	17	17	18	18	19	19
27	15	16	16	17	17	18	18	18	19	19	20
28	16	16	17	17	18	18	19	19	20	20	21
29	16	17	17	18	18	19	19	20	20	21	21
30	17	17	18	18	19	19	20	20	21	21	22
Mos.											
1	17	17	18	18	19	19	20	20	21	21	22
2	34	35	36	37	38	39	40	41	42	43	44
3	51	53	54	56	57	59	60	62	63	65	66
4	68	70	72	74	76	78	80	82	84	86	88
5	85	88	90	93	95	98	100	103	105	108	110
6	102	105	108	111	114	117	120	123	126	129	132
7	119	123	126	130	133	137	140	144	147	151	154
8	136	140	144	148	152	156	160	164	168	172	176
9	153	158	162	167	171	176	180	185	189	194	198
10	170	175	180	185	190	195	200	205	210	215	220
11	187	193	198	204	209	215	220	226	231	237	242
12	204	210	216	222	228	234	240	246	252	258	264

Interest Table at 6 Per Cent.—*Continued.*

Days.	DOLLARS.										
	45	46	47	48	49	50	51	52	53	54	55
1	1	1	1	1	1	1	1	1	1	1	1
2	1	2	2	2	2	2	2	2	2	2	2
3	2	2	2	2	2	2	3	3	3	3	3
4	3	3	3	3	3	3	3	3	3	4	4
5	4	4	4	4	4	4	4	4	4	4	4
6	4	5	5	5	5	5	5	5	5	5	5
7	5	5	5	6	6	6	6	6	6	6	6
8	6	6	6	6	6	7	7	7	7	7	7
9	7	7	7	7	7	7	8	8	8	8	8
10	7	8	8	8	8	8	8	9	9	9	9
11	8	8	8	9	9	9	9	9	10	10	10
12	9	9	9	9	10	10	10	10	10	11	11
13	10	10	10	10	10	11	11	11	11	12	12
14	11	11	11	11	11	12	12	12	12	12	13
15	11	12	12	12	12	12	13	13	13	13	14
16	12	12	12	13	13	13	13	14	14	14	14
17	13	13	13	14	14	14	14	15	15	15	15
18	13	14	14	14	14	15	15	15	16	16	16
19	14	15	15	15	15	16	16	16	17	17	17
20	15	15	16	16	16	16	17	17	17	18	18
21	16	16	16	17	17	18	18	18	19	19	19
22	16	17	17	18	18	18	19	19	19	20	20
23	17	18	18	18	19	19	20	20	20	21	21
24	18	18	19	19	20	20	20	21	21	22	22
25	19	19	20	20	20	21	21	22	22	23	23
26	20	20	20	21	21	22	22	23	23	23	24
27	20	21	21	22	22	22	23	23	24	24	25
28	21	21	22	22	23	23	24	24	25	25	26
29	22	22	23	23	24	24	25	25	26	26	27
30	22	23	23	24	24	25	25	26	26	27	27
Mos.											
1	22	23	23	24	24	25	25	26	26	27	27
2	45	46	47	48	49	50	51	52	53	54	55
3	68	69	71	72	74	75	77	78	80	81	83
4	90	92	94	96	98	100	102	104	106	108	110
5	113	115	118	120	123	125	128	130	133	135	138
6	135	138	141	144	147	150	153	156	159	162	165
7	158	161	165	168	172	175	179	182	186	189	193
8	180	184	188	192	196	200	204	208	212	216	220
9	203	207	212	216	221	225	230	234	239	243	248
10	225	230	235	240	245	250	255	260	265	270	275
11	248	253	259	264	270	275	281	286	292	297	303
12	270	276	282	288	294	300	306	312	318	324	330

Interest Table at 6 Per Cent.—*Continued.*

Days.	DOLLARS.										
	56	57	58	59	60	61	62	63	64	65	66
1	1	1	1	1	1	1	1	1	1	1	1
2	2	2	2	2	2	2	2	2	2	2	2
3	3	3	3	3	3	3	3	3	3	3	3
4	4	4	4	4	4	4	4	4	4	4	4
5	5	5	5	5	5	5	5	5	5	5	5
6	6	6	6	6	6	6	6	6	6	6	7
7	6	7	7	7	7	7	7	7	7	7	8
8	7	7	8	8	8	8	8	8	8	9	9
9	8	8	9	9	9	9	9	9	9	10	10
10	9	9	10	10	10	10	10	10	11	11	11
11	10	10	10	11	11	11	11	11	12	12	12
12	11	11	11	12	12	12	12	12	13	13	13
13	12	12	12	13	13	13	13	13	14	14	14
14	13	13	13	14	14	14	14	14	15	15	15
15	14	14	14	15	15	15	15	16	16	16	16
16	15	15	15	16	16	16	16	17	17	17	17
17	16	16	16	16	17	17	17	18	18	18	18
18	17	17	17	17	18	18	18	19	19	19	20
19	17	18	18	18	19	19	19	20	20	20	21
20	18	19	19	19	20	20	20	21	21	21	22
21	20	20	20	21	21	21	22	22	22	23	23
22	20	21	21	22	22	22	23	23	23	24	24
23	21	22	22	23	23	23	24	24	25	25	25
24	22	23	23	24	24	24	25	25	26	26	26
25	23	24	24	25	25	25	26	26	27	27	27
26	24	25	25	26	26	26	27	27	28	28	29
27	25	26	26	27	27	27	28	28	29	29	30
28	26	27	27	28	28	28	29	29	30	30	31
29	27	28	28	29	29	29	30	30	31	31	32
30	28	28	29	29	30	30	31	31	32	32	33
Mos.											
1	28	28	29	29	30	30	31	31	32	32	33
2	56	57	58	59	60	61	62	63	64	65	66
3	81	86	87	89	90	92	93	95	96	98	99
4	112	114	116	118	120	122	124	126	128	130	132
5	140	143	145	148	150	153	155	158	160	163	165
6	168	171	174	177	180	183	186	189	192	195	198
7	196	200	203	207	210	214	217	221	224	228	231
8	224	228	232	236	240	244	248	252	256	260	264
9	252	257	261	266	270	275	279	284	288	293	297
10	280	285	290	295	300	305	310	315	320	325	330
11	308	314	319	325	330	336	341	347	352	358	363
12	336	342	348	354	360	366	372	378	384	390	396

Interest Table at 6 Per Cent.—*Continued.*

Days.	67	68	69	70	71	72	73	74	75	76	77
1	1	1	1	1	1	1	1	1	1	1	1
2	2	2	2	2	2	2	2	2	2	2	3
3	3	3	3	3	4	4	4	4	4	4	4
4	4	4	5	5	5	5	5	5	5	5	5
5	6	6	6	6	6	6	6	6	6	6	6
6	7	7	7	7	7	7	7	7	7	7	8
7	8	8	8	8	8	8	8	9	9	9	9
8	9	9	9	9	9	9	10	10	10	10	10
9	10	10	10	10	11	11	11	11	11	11	11
10	11	11	11	12	12	12	12	12	12	12	13
11	12	12	12	13	13	13	13	13	14	14	14
12	13	13	14	14	14	14	14	15	15	15	15
13	14	15	15	15	15	15	16	16	16	16	16
14	15	16	16	16	16	17	17	17	17	17	18
15	17	17	17	17	18	18	18	18	18	19	19
16	18	18	18	18	19	19	19	19	20	20	20
17	19	19	19	20	20	20	20	21	21	21	22
18	20	20	20	21	21	21	22	22	22	22	23
19	21	21	22	22	22	22	23	23	23	24	24
20	22	22	23	23	23	24	24	24	25	25	25
21	23	24	24	25	25	25	26	26	26	27	27
22	25	25	25	26	26	26	27	27	27	28	28
23	26	26	27	27	27	28	28	28	29	29	30
24	27	27	28	28	28	29	29	30	30	30	31
25	28	28	29	29	30	30	30	31	31	32	32
26	29	29	30	30	31	31	32	32	32	33	33
27	30	31	31	32	32	32	33	33	34	34	35
28	31	32	32	33	33	34	34	35	35	35	36
29	32	33	33	34	34	35	35	36	36	37	37
30	33	34	34	35	35	36	36	37	37	38	38
Mos.											
1	33	34	34	35	35	36	36	37	37	38	38
2	67	68	69	70	71	72	73	74	75	76	77
3	101	102	104	105	107	108	110	111	113	114	116
4	134	136	138	140	142	144	146	148	150	152	154
5	168	170	173	175	178	180	183	185	188	190	193
6	201	204	207	210	218	216	219	222	225	228	231
7	235	238	242	245	249	252	256	259	263	266	270
8	268	272	276	280	284	288	292	296	300	304	308
9	302	306	311	315	320	324	329	333	338	342	347
10	335	340	345	350	355	360	365	370	375	380	385
11	369	374	380	385	391	396	402	407	413	418	424
12	402	408	414	420	426	432	438	444	450	456	462

F

Interest Table at 6 Per Cent.—*Continued.*

Days.	78	79	80	81	82	83	84	85	86	87	88
1	1	1	1	1	1	1	1	1	1	1	1
2	3	3	3	3	3	3	3	3	3	3	3
3	4	4	4	4	4	4	4	4	4	4	4
4	5	5	5	5	5	5	6	6	6	6	6
5	6	6	7	7	7	7	7	7	7	7	7
6	8	8	8	8	8	8	8	8	8	8	9
7	9	9	9	9	9	10	10	10	10	10	10
8	10	10	11	11	11	11	11	11	11	11	12
9	12	12	12	12	12	12	12	13	13	13	13
10	13	13	13	13	13	14	14	11	14	11	14
11	14	14	14	15	15	15	15	15	16	16	16
12	15	16	16	16	16	16	17	17	17	17	17
13	17	17	17	17	18	18	18	18	18	19	19
14	18	18	18	19	19	19	19	20	20	20	20
15	19	19	20	20	20	20	21	21	21	21	22
16	21	21	21	21	22	22	22	22	23	23	23
17	22	22	22	23	23	23	23	24	24	24	25
18	23	23	24	24	24	25	25	25	25	26	26
19	24	25	25	25	26	26	26	27	27	27	27
20	26	26	26	27	27	27	28	28	28	29	29
21	27	28	28	28	29	29	29	30	30	30	31
22	29	29	29	30	30	30	31	31	32	32	33
23	30	30	31	31	31	32	32	33	33	33	34
24	31	32	32	32	33	33	34	34	34	35	35
25	32	33	33	34	34	35	35	35	36	36	37
26	34	34	35	35	36	36	36	37	37	38	38
27	35	36	36	36	37	37	38	38	39	39	40
28	36	37	37	38	38	39	39	40	40	41	41
29	38	38	39	39	40	40	41	41	42	42	43
30	39	39	40	40	41	41	42	42	43	43	44
Mos.											
1	39	39	40	40	41	41	42	42	43	43	44
2	78	79	80	81	82	83	84	85	86	87	88
3	117	119	120	122	123	125	126	128	129	131	132
4	156	158	160	162	164	166	168	170	172	174	176
5	195	198	200	203	205	208	210	213	215	218	220
6	234	237	240	243	246	249	252	255	258	261	264
7	273	277	280	284	287	291	294	298	301	305	308
8	312	316	320	324	328	332	336	340	344	348	352
9	351	356	360	365	369	374	378	383	387	392	396
10	390	395	400	405	410	415	420	425	430	435	440
11	429	435	440	446	451	457	462	468	473	479	484
12	468	474	480	486	492	498	504	510	516	522	528

Interest Table at 6 Per Cent.—*Continued.*

Days.	89	90	91	92	93	94	95	96	97	98	99
1	1	1	1	2	2	2	2	2	2	2	2
2	3	3	3	3	3	3	3	3	3	3	3
3	4	4	4	5	5	5	5	5	5	5	5
4	6	6	6	6	6	6	6	6	6	6	7
5	7	7	7	8	8	8	8	8	8	8	8
6	9	9	9	9	9	9	9	10	10	10	10
7	10	10	10	11	11	11	11	11	11	11	12
8	12	12	12	12	12	12	13	13	13	13	13
9	13	13	13	14	14	14	14	14	15	15	15
10	15	15	15	15	15	15	16	16	16	16	16
11	16	16	16	17	17	17	17	18	18	18	18
12	18	18	18	18	19	19	19	19	19	20	20
13	19	19	19	20	20	20	21	21	21	21	21
14	20	21	21	21	22	22	22	22	23	23	23
15	22	22	22	23	23	23	24	24	24	24	25
16	23	24	24	24	25	25	25	26	26	26	26
17	25	25	25	26	26	27	27	27	27	28	28
18	26	27	27	27	28	28	29	29	29	29	30
19	28	28	28	29	29	30	30	30	31	31	31
20	29	30	30	30	31	31	32	32	32	33	33
21	31	31	32	32	33	33	33	34	34	34	35
22	33	33	33	34	34	34	35	35	36	36	36
23	34	35	35	35	36	36	36	37	37	38	38
24	36	36	36	37	37	38	38	38	39	39	40
25	37	38	38	38	39	39	40	40	40	41	41
26	39	39	39	40	40	41	41	42	42	42	43
27	40	40	41	41	42	42	43	43	44	44	45
28	42	42	42	43	43	44	44	45	45	46	46
29	43	44	44	44	45	45	46	46	47	47	48
30	44	45	45	46	46	47	47	48	48	49	49
Mos.											
1	44	45	45	46	46	47	47	48	48	49	49
2	89	90	91	92	93	94	95	96	97	98	99
3	134	135	137	138	139	141	142	144	145	147	148
4	178	180	182	184	186	188	190	192	194	196	198
5	223	225	228	230	232	235	237	240	242	245	247
6	267	270	273	276	279	282	285	288	291	294	297
7	312	315	319	322	325	329	332	336	339	343	346
8	356	360	364	368	372	376	380	384	388	392	396
9	401	405	410	414	418	423	427	432	436	441	445
10	445	450	455	460	465	470	475	480	485	490	495
11	490	495	501	506	511	517	522	528	533	539	544
12	534	540	546	552	558	564	570	576	582	588	594

Interest Table at 6 Per Cent.—*Continued.*

Days.	100	200	300	400	500	600	700	800	900	1000	2000
1	2	3	5	7	8	10	12	13	15	16	33
2	3	7	10	13	17	20	23	27	30	33	66
3	5	10	15	20	25	30	35	40	45	50	100
4	7	13	20	27	33	40	47	53	60	67	133
5	8	17	25	33	42	50	58	67	75	83	167
6	10	20	30	40	50	60	70	80	90	100	200
7	12	23	35	47	58	70	82	93	105	117	233
8	13	26	40	53	67	80	93	107	120	133	267
9	15	30	45	60	75	90	105	120	135	150	300
10	17	33	50	67	83	100	117	133	150	167	333
11	18	37	55	73	92	110	128	147	165	183	367
12	20	40	60	80	100	120	140	160	180	200	400
13	22	43	65	87	108	130	152	173	195	217	433
14	23	47	70	93	117	140	163	187	210	233	467
15	25	50	75	100	125	150	175	200	225	250	500
16	27	53	80	107	133	160	187	213	240	267	533
17	28	57	85	113	142	170	198	227	255	283	567
18	30	60	90	120	150	180	210	240	270	300	600
19	32	63	95	127	158	190	222	253	285	317	633
20	33	67	100	133	167	200	233	267	300	333	667
21	35	70	105	140	175	210	245	280	315	350	700
22	37	73	110	147	183	220	257	293	330	367	733
23	38	77	115	153	192	230	268	307	345	383	767
24	40	80	120	160	200	240	280	320	360	400	800
25	42	83	125	167	208	250	292	333	375	417	833
26	43	87	130	173	217	260	303	347	390	433	867
27	45	90	135	180	225	270	315	360	405	470	900
28	47	93	140	187	233	280	327	373	420	467	933
29	48	97	145	193	242	290	338	387	435	483	967
30	50	100	150	200	250	300	350	400	450	500	1000
Mos.											
1	50	100	150	200	250	300	350	400	450	500	1000
2	100	200	300	400	500	600	700	800	900	1000	2000
3	150	300	450	600	750	900	1050	1200	1350	1500	3000
4	200	400	600	800	1000	1200	1400	1600	1800	2000	4000
5	250	500	750	1000	1250	1500	1750	2000	2250	2500	5000
6	300	600	900	1200	1500	1800	2100	2400	2700	3000	6000
7	350	700	1050	1400	1750	2100	2450	2800	3150	3500	7000
8	400	800	1200	1600	2000	2400	2800	3200	3600	4000	8000
9	450	900	1350	1800	2250	2700	3150	3600	4050	4500	9000
10	500	1000	1500	2000	2500	3000	3500	4000	4500	5000	10000
11	550	1100	1650	2200	2750	3300	3850	4400	4950	5500	11000
12	600	1200	1800	2400	3000	3600	4200	4800	5400	6000	12000

Compound Interest by Decimals, at 7 Per Cent., from One Year to Ten Years.

Principal, 100 cents.

1.07 — **1** year.
1.1449 — **2** years.
1.225043 — **3** years.
1.31079601 — **4** years.
1.4025517307 — **5** years.
1.500730351849 — **6** years.
1.60578147647843 — **7** years.
1.7181861798319201 — **8** years.
1.828459212420154507 — **9** years.
2.01130513366216995770 — **10** years.

Table of Compound Interest at 6 Per Cent.

Principal, 100 cents.

1.06 — **1** year.
1.1236 — **2** years.
1.191016 — **3** years.
1.26247696 — **4** years.
1.3382255776 — **5** years.
1.418519112256 — **6** years.
1.50363025899136 — **7** years.
1.5938480745308416 — **8** years.
1.689478959002692096 — **9** years.
1.79084769654285362176 — **10** years.
1.8982985583354248390656 — **11** years.
2.012196498355503206409536 — **12** years.

USE OF THE ABOVE TABLES.— Required the compound interest of $100, for 10 years, at 6 per cent.

ANS.—$79.08.4 mills.

1.7908476   tabular number.
      100 × by the principal.

179.0847600
100   subtract the principal.

79.08.4

8

Cent Table at 6 Per Cent.

Days	1	2	3	4	5	6	7	8	9	10	11	12	24	36
50	0	0	1	1	1	1	2	2	2	3	3	3	6	9
49	0	0	1	1	1	1	2	2	2	3	3	3	6	9
48	0	0	1	1	1	1	2	2	2	2	3	3	6	9
46 47	0	0	1	1	1	1	2	2	2	2	3	3	6	9
44 45	0	0	1	1	1	1	2	2	2	2	2	3	5	8
42 43	0	0	1	1	1	1	2	2	2	2	2	3	5	8
40 41	0	0	1	1	1	1	2	2	2	2	2	2	5	7
39	0	0	1	1	1	1	2	2	2	2	2	2	5	7
38	0	0	0	1	1	1	2	2	2	2	2	2	5	7
37	0	0	0	0	1	1	2	2	2	2	2	2	5	7
36	0	0	0	0	1	1	1	2	2	2	2	2	4	6
34 35	0	0	0	0	1	1	1	1	2	2	2	2	4	6
33 32	0	0	0	0	1	1	1	1	2	2	2	2	4	6
31	0	0	0	0	1	1	1	1	2	2	2	2	4	6
30	0	0	0	0	1	1	1	1	2	2	2	2	4	5
28 29	0	0	0	0	1	1	1	1	2	2	2	2	4	5
27	0	0	0	0	1	1	1	1	1	1	2	3	5	
26	0	0	0	0	1	1	1	1	1	1	2	3	5	
25	0	0	0	0	1	1	1	1	1	1	2	3	5	
23 24	0	0	0	0	1	1	1	1	1	1	1	3	4	
21 22	0	0	0	0	1	1	1	1	1	1	1	3	4	
19 20	0	0	0	0	0	1	1	1	1	1	1	3	4	
18	0	0	0	0	0	1	1	1	1	1	1	2	3	
17	0	0	0	0	0	1	1	1	1	1	1	2	3	
16	0	0	0	0	0	0	1	1	1	1	1	2	3	
15	0	0	0	0	0	0	1	1	1	1	1	2	3	
14	0	0	0	0	0	0	0	1	1	1	1	2	3	
13	0	0	0	0	0	0	0	1	1	1	1	2	3	
12	0	0	0	0	0	0	0	0	1	1	1	1	2	
11	0	0	0	0	0	0	0	0	0	1	1	1	2	
10	0	0	0	0	0	0	0	0	0	1	1	1	2	
9	0	0	0	0	0	0	0	0	0	0	1	1	2	
7 8	0	0	0	0	0	0	0	0	0	0	0	0	1	2
5 6	0	0	0	0	0	0	0	0	0	0	0	0	1	1
3 4	0	0	0	0	0	0	0	0	0	0	0	0	0	1

(cents)	1	2	3	4	5	6	7	8	9	10	11	12	24	36
99	0	1	1	2	2	3	3	4	4	5	5	6	12	18
97 98	0	1	1	2	2	3	3	4	4	5	5	6	12	18
96	0	1	1	2	2	3	3	4	4	5	5	6	12	17
95	0	1	1	2	2	3	3	4	4	5	5	6	11	17
93 94	0	1	1	2	2	3	3	4	4	5	5	6	11	17
92	0	1	1	2	2	3	3	4	4	5	5	6	11	17
90 91	0	1	1	2	2	3	3	4	4	5	5	5	11	16
88 89	0	1	1	2	2	3	3	4	4	4	5	5	11	16
87	0	1	1	2	2	3	3	3	4	4	5	5	10	16
86	0	1	1	2	2	3	3	3	4	4	5	5	10	15
84 85	0	1	1	2	2	3	3	3	4	4	5	5	10	15
82 83	0	1	1	2	2	2	3	3	4	4	5	5	10	15
81	0	1	1	2	2	2	3	3	4	4	4	5	10	15
80	0	1	1	2	2	2	3	3	4	4	4	5	10	14
78 79	0	1	1	2	2	2	3	3	4	4	4	5	9	14
77	0	1	1	2	2	2	3	3	3	4	4	5	9	14
75 76	0	1	1	2	2	2	3	3	3	4	4	5	9	14
74	0	1	1	1	2	2	3	3	3	4	4	4	9	13
72 73	0	1	1	1	2	2	3	3	3	4	4	4	9	13
71	0	1	1	1	2	2	2	3	3	4	4	4	9	13
70	0	1	1	1	2	2	2	3	3	4	4	4	8	13
68 69	0	1	1	1	2	2	2	3	3	3	4	4	8	12
66 67	0	1	1	1	2	2	2	3	3	3	4	4	8	12
64 65	0	1	1	1	2	2	2	3	3	3	4	4	8	12
63	0	1	1	1	2	2	2	3	3	3	3	4	8	11
61 62	0	1	1	1	2	2	2	2	3	3	3	4	7	11
59 60	0	1	1	1	2	2	2	2	3	3	3	4	7	11
58	0	1	1	1	1	2	2	2	3	3	3	3	7	10
57	0	1	1	1	1	2	2	2	3	3	3	3	7	10
55 56	0	1	1	1	1	2	2	2	3	3	3	3	7	10
53 54	0	1	1	1	1	2	2	2	2	3	3	3	6	10
51 52	0	1	1	1	1	2	2	2	2	3	3	3	6	9

EXPLANATION OF THE TABLE.—The cents are given at the heads of the pages, 3 to 99, and the months on the left.

EXAMPLE.—Interest on 39 cents: 9 months, 2 cents. On 65 cents: 11 months, 3 cents. On 92 cents: 9 months, 4 cents.

Simple Interest.

What is the interest on $942.29, for one year, at 6 per cent.?

$$100 : 6 :: 942.29$$
$$6$$
$$100) 5653.74 (56.5374 = \$56.53\tfrac{74}{100}.$$

Rules for Calculating Interest.

To find the interest on any principal for any number of days, the answer in each case being in cents, separate the two right hand figures of answer to express in dollars and cents. Rule as follows:

For 4 per cent.— Multiply the principal by the number of days to run, separate the right hand figures from the product, and divide by 9.

For 5 per cent.— Multiply by the number of days, and divide by 72.

For 6 per cent.— Multiply by the number of days, separate right hand figure, and divide by 6.

For 8 per cent.— Multiply by the number of days, and divide by 45.

For 9 per cent.— Multiply by the number of days, separate right hand figure, and divide by 4.

For 10 per cent.— Multiply by the number of days, and divide by 36.

For 12 per cent.— Multiply by the number of days, separate right hand figure, and divide by 3.

For 15 per cent.— Multiply by the number of days, and divide by 24.

For 18 per cent.— Multiply by the number of days, separate the right-hand figure, and divide by 2.

For 20 per cent.— Multiply by the number of days, and divide by 18.

For 24 per cent.— Multiply by the number of days, and divide by 15.

A Table Showing the Amount of $100, at Compound Interest, at 6 Per Cent.

1 year	. . . $106.00	7 years .	. . $150.36
2 years	. . . 112.36	8 "	. . . 159.38
3 "	. . . 119.10	9 "	. . . 168.95
4 "	. . . 126.24	10 "	. . . 179.08
5 "	. . . 133.82	11 "	. . . 189.82
6 "	. . . 141.85	12 "	. . . 201.21

Second term of 12 years	402
Third " " " "	804
Fourth " " " "	1608
Fifth " " " "	3216
Sixth " " " "	6432
Seventh " " " "	12864
Eighth " " " "	25728
Ninth " " " "	51456
Tenth " " " "	102912
Eleventh " " " "	205824
Twelfth " " " "	411648

and at this rate would in two hundred years amount to 13 millions, and in three hundred years, to about 4 thousand millions of dollars.

If I purchase of National bank-stock, that produces a half-yearly dividend of 4½ per cent., 100 shares, par $50 each, $5000, at 30 per cent. advance, for which I pay in cash $6500, required how much per cent. per annum I get on the money advanced.

Interest on the stock for 12 months, $450.00

Interest on the interest for 6 months,   10.12½

So the $5000 stock, or $6500 cash, has produced $460.12½ in one year.

Now say, as the money the stock cost, $6500, is to the interest of the stock for one year, $460.12½, so is $100 to its

8 *

interest for one year, which gives $7.079 per cent. per annum nearly.

PROOF.—Multiply $6.500 by $7.079, which equals $460.13.

Discount

Is the deducting of a sum for the prompt or advanced payment of obligations, notes, etc. Most persons deduct the rate per cent. from the sum for the discount; banks also.

Although practice makes it common, this system is contrary to equity.

The sum deducted should be such that the balance at the rate per cent. given would amount to the original.

The rule is as 100, with the rate per cent. added, is to the rate per cent. so is the sum given to the discount; that is, $106 : 600 :: 100 : 566\frac{4}{106}$, or × by 3 ÷ 53.

In the other case, if $6 are deducted, the charge is more than 6 per cent.

Table of the Approximate Value of Foreign Coins in United States Currency.

Augustus, Saxony...........	$3.98	Dollar, Norway, Sweden,	
Carolin, Bavaria............	4.93	Spain $1.05	
Copeck, Russia..............	34	Dollar, Rigsbank, Denmark,	.52
Crown, Baden, Bavaria....	1.06	Dollar, Specie-Dollar, Den-	
Crown, or 5s., English.....	1.13	mark............................	1.05
Crown, Portugal............	1.12	Doubloon, Central Amer-	
Crown, Sardinia............	.92	ica..................$14.50 to 15.65	
Crown, Sicily................	.96	Doubloon, New Granada....	15.34
Crown, Spanish, half pis-		Doubloon, Spain, Mexico....	15.65
tole.........................	1.95	Drachm, Greece..............	.18
Dollar, Central America..	.95	Ducat, Austria, Bohemia,	
Dollar, Chili, Mexico,		Hamburg, Hanover......	2.28
Peru, Bolivia, New Gra-		Ducat, Denmark..............	1.81
nada......................	1.00	Ducat, Sweden...............	2.20

Ducatoon, Holland	$1.32	Pound, English	$4.84
Florin, Austria, Silesia	.48	Pistareen, Spain	.20
Florin, (gold) Hanover	1.66	Piastre, Egypt	.05
Florin, (silver) Hanover	.56	Piastre, Spain	1.04
Florin, Holland, Netherlands, South Germany	.40	Piastre, Turkey, old	.42
		Piastre, Turkey, new	.04$\frac{2}{10}$
Florin, Prussian	.55	Pistola, Rome	3.37
Franc, France and Belgium	.19	Pistole, Spain	3.90
5-franc piece	.95	Reale, Central America, average	.05¾
Guilder, British Guiana	.26	Reale Plate, Spain	.10
Guilder, Netherlands	.40	Reale Vellon, Spain	.05
Guinea, 21s., English	5.08	2 Reales, Ecuador	.18¾
Gulden, Baden	.40	Reis (1200), Brazil	.99
Groschen, Poland, Prussia	.02$\frac{1}{10}$	Rix-Dollar, Baden, Brunswick	1.00
5 Groschen	.12		
Grote, Bremen	.01	Rix-Dollar, Bavaria, Austria, Hungary	.97
Imperial, Russia	7.92		
10 Kreutzer, Austria	.08	Rix-Dollar, Hanover	1.10
60 Kreutzer, or Florin, Austria	.48	Rix-Dollar, Sweden and Holland	1.05
Kreutzer, Bavaria	⅔	Rouble, Russia	.79
Lira, Milan	.14	5 Roubles, Russia	3.95
Livre, France, Sardinia	.18½	Rupee	.53
Livre, Tuscany, Venice	.16	Rupee, Bombay, Madras	.45
Marc, Denmark	.09		
Maximilian, Bavaria	3.30	Schilling, Hamburg	.02
Milrea, Portugal	1.12	Scudo, Genoa	1.28
Mohur, Bengal	8.15	Scudo, Naples, Sicily	.95
Mohur, Bombay	7.28	Scudo, Piedmont	1.36
Moidore, Portugal	6.50	Scudo, Rome	1.00
Napoleon, 20 Francs, France	3.84	Scudo, Sardinia	.92
Ounce, Sicily	2.50	Shilling, English	.23
Paolo, Rome	.10	Skilling, Denmark	¾
Para	$\frac{8}{10}$	Sou, France, nearly	.01
Peseta, Spain	.20	Sovereign, English	4.84
Pound, Canada, New Brunswick, etc	4.00	Sovrano, Austria, Bohemia	3.57
		Specie thaler, Saxony	.98

Star, Pagoda, Madras........ $1.81	men, Saxony, Hanover,	
Stiver, Holland, nearly...... .02	Poland......... .... $ .69	
Teston, Rome .... .30	10 thaler, Prussia............. 8.00	
Testoon, Portugal............. .12	Zecchin, Turkey............... 1.40	
Thaler, N. Germany, Bre-	Zecchino, Rome............... 2.27	

The United States coins are: Copper — ½, 1, and 2 cents. Nickel and copper — 1, 3, and 5 cents. Nickel and silver — 3 cents. Silver — half-dime = 5 cents; one dime = 10 cents; two dimes = 20 cents; quarter-dollar = 25 cents; half-dollar = 50 cents; legal-tender dollar, 412½ grains; trade-dollar, 420 grains. Gold—1 and 3 dollars; quarter-eagle = 2.50; half-eagle = 5.00; eagle = 10.00, and double eagle = 20.00.

The denomination of U. S. money, called Federal money, from the confederation of the States, was established by act of Congress in the year 1785, and are eagle, dollar, dime, cent, and mill.

Rule to reduce or change Canada currency to Federal money.— Reduce to pence and divide by 60.

Rule to reduce British Sterling to dollars and cents.—Reduce it to 6 pence and divide by 9, because nine sixpences make one dollar, and $4.44.4 U. S. money — £1 sterling.

Example.—

$$
\begin{array}{ccc}
£ & s. & d. \\
1 & 0 & 0 \\
20 & & \\
\overline{20} & & \\
2 & & \\
\hline
9\,)\,40 & & \\
\hline
\$4.44.4 & &
\end{array}
$$

4 farthings = 1 penny, 12 pence = 1 shilling, and 20 shillings = one £.

To reduce Sterling to dollars and cents, valuing the pound, with exchange added, at $4.80 to the given pound. — Annex the shillings and pence in decimals of a pound, and add a cipher, or move the decimal point one place to the right; then multiply by 8 and the product by 6, and the last product is the value in dollars and cents.

To reduce or change Federal money to Sterling or English money.

EXAMPLE.— Multiply the cents by 54, cut off two figures to the right, and you have pence.

EXAMPLE.—

$$
\begin{array}{r}
\$4.44.4 \\
54 \\
\hline
1776 \\
2220 \\
24 \\
\hline
12\,)\,2400\cancel{0}\cancel{0} \\
\hline
2\cancel{0}\,)\,2\cancel{0} \\
\hline
\pounds1
\end{array}
$$

To reduce dollars and cents to pounds at the rate of $4.80 to the pound.—Divide the given sum by 8 and the last quotient by 6; placing the decimal point one place to the left gives the required sum in pounds and decimal parts.

To reduce any sum of French francs and centimes to dollars and cents.—Take $\frac{1}{8}$, and $\frac{1}{4}$ of this $\frac{1}{8}$, and add the two together for the value required in dollars and cents.

To reduce any sum in dollars and cents to francs and centimes. — Multiply the given sum by 4, and to this product add one-third of it for the value required.

Insurance.

Insurance is a premium or percentage paid to a company for insuring against losses by fire or transportation. The company or underwriters issue to the insurer a policy of insurance, which contains the conditions of the contract; the losses are regulated by the amount of damage, and paid accordingly.

If the damage is on land, settlements are made without much delay.

If the loss is on the sea, there is sometimes considerable delay in the settlements, which are called averages.

Trading vessels are generally held on shares. The manager, or ship's-husband, being part owner, is allowed a percentage on the amount of freight. The settlements are made the same as other partnership concerns; but in the settlement of losses, an important operation frequently occurs in settling averages, which are classed as general and particular, also called gross and simple.

Averages.

General average is a proportion paid by the proprietors for losses that are made with a view to safety — cutting away masts, etc., and throwing cargo overboard.

Particular average is a contribution for such damages as may happen from common accidents at sea. This is paid by the proprietors of the articles which suffered the damage, and the calculations are made according to the rules of Fellowship.

In computing general average for cutting masts, etc., a deduction of one-third is made from the cost, as new articles are presumed to be that much better; but goods thrown overboard are valued at the sum they would have netted had they arrived safely.

When a ship's cargo and freight are fully insured, the underwriters are responsible to the proprietors for both general and particular averages.

EXAMPLE.—Suppose the ship Grey Eagle, from the West Indies to Philadelphia, in the course of her voyage, to have suffered the following damages: required the general and particular averages.

General Average.

Cost of replacing masts, cables, etc., cut away, $1500		
Deduct one-third for newness, . . . 500		1000
Anchor lost, which cost		200
10 puncheons of rum thrown overboard . . .		1000
Sundry charges for pilotage, etc., ...		200
Amount of general loss . .		$2400

Particular Average.

Of 80 hhds. of sugar shipped, a part was so much damaged that the deficiency in 20, on a comparison with 60 that arrived safely, was 10 hhds., which at $120 each amounted to $1200.

SHIP, CARGO, AND FREIGHT.

Grey Eagle valued at $22000		
Cargo, net proceeds 34000		
Gross freight 10000		
Portage bill deducted 200		9800
Total amount $65800		

Statement of General Average.
As 65,800 : 2,400 :: 100 : 3.65.

Statement for Particular Average.
As the value of 80 hhds. of sugar, which is 9,600 : 1,200 :: 100 : 12.50.

So the underwriters or insurers will have to pay 3.65 per cent., nearly, for general average on $65800, and 12.50 per cent. for particular average on $6600.

Fellowship.

Is a method of showing the gains or losses of parties in joint operations.

RULE.—As the whole stock is to the whole gain or loss so is each share to the gain or loss.

EXAMPLE.—Three persons purchase a factory for $100,000, A agreeing to subscribe $10,000; B, $40,000; C, $50,000, after the purchase. An offer is made by D, of $150,000, and accepted by them. Required the share of each in the profit.

$$100000 : 50000 :: 10000 :\ 5,000,\ \text{A's share.}$$
$$100000 : 50000 :: 40000 : 20,000,\ \text{B's "}$$
$$100000 : 50000 :: 50000 : 25,000,\ \text{C's "}$$

Suppose they subscribe the amount, and manufacture goods for one year, and their profits net $50,000 at the end of the year. Required each one's capital.

$$100000 : 150000 :: 10000 : 15,000\ \text{A's capital.}$$
$$100000 : 150000 :: 40000 : 60,000\ \text{B's "}$$
$$100000 : 150000 :: 50000 : 75,000\ \text{C's "}$$

DOUBLE FELLOWSHIP, WITH TIME.—RULE.—Multiply each share by the time of its interest in the fellowship; then as the sum of the products is to the product of each interest so is the whole gain or loss to each share of the gain or loss.

EXAMPLE.—A ship's company take a prize worth $10,000, which they divide according to their rate of pay and time of service on board.

The officers have been on board 6 months and the men 3 months. The pay of the lieutenants is $100, midshipmen $50, and men $10 per month; and there are 2 lieutenants, 4 midshipmen, and 50 men. What is each one's share?

$$
\begin{aligned}
\text{2 lieutenants,} \quad \$100 &= 200 \times 6 = 1200 \\
\text{4 midshipmen,} \quad \$50 &= 200 \times 6 = 1200 \\
\text{50 men,} \quad \$10 &= 500 \times 3 = \underline{1500} \\
&\qquad\qquad\qquad\quad 3900
\end{aligned}
$$

Lieutenants, 3900 : 1200 :: 10000 : 3,076.92 ÷ 2 = $1,538.46
Midshipmen, 3900 : 1200 :: 10000 : 3,076.92 ÷ 4 = 769.23
Men,   3900 : 1500 :: 10000 : 3,846.16 ÷ 50 = 76.92

Weights and Measures.

Diamond Weight.

| 16 parts | = | 1 grain | = | 0.8 Troy grains. |
| 4 grains | = | 1 carat | = | 3.2 " " |

The grain of diamond weight is equal to $\frac{8}{10}$ grain of Troy Weight.

The carat equals $3\frac{1}{5}$ grains of Troy Weight.

Articles Sold by Number.

12 units	=	1 dozen,
12 dozen, or 144 units,	=	1 gross,
12 gross, or 1728 units,	=	1 great gross,
20 units	=	1 score.

Paper Measures.

24 sheets	=	1 quire,
20 quires, or 480 sheets,	=	1 ream,
2 reams, or 960 sheets,	=	1 bundle.

Cloth Measures.

2¼ inches	=	1 nail,
4 nails	=	1 quarter,
4 quarters	=	1 yard.

Cubic Measures.

| 1728 cubic inches | = | 1 cubic foot, |
| 27 " feet | = | 1 " yard. |

9   G

Book Measures.

2 leaves are 1 folio,
4 "   " 1 quarto,
8 "   " 1 octavo,
12 "   " 1 duodecimo.

To all sizes above these add mo, as, 16mo, 18mo, 24mo, etc.

The standard measure of length for the United States and England is, theoretically, that of a pendulum vibrating seconds in the latitude of London, at a temperature of 62° Fah., in a vacuum, and at the level of the sea. The length of the pendulum was originally supposed to be divided into 39.1393 equal parts or inches, of which 36 inches were adopted as the standard yard. The original standard was lost by fire in London, and could not be restored by pendulum, and at present the British yard measure is shorter than that of the United States by $\frac{1}{14}$ of an inch in 100 feet.

Apothecaries' Weights.

			Grains.	Scruples.	Drams.
20 grains	=	1 scruple,			
3 scruples	=	1 dram	=	60	
8 drams	=	1 ounce	=	480 =	24
12 ounces	=	1 pound	= 5760	= 288	= 96

In Troy and Apothecaries' Weights, the grain, ounce, and pound are the same.

Troy Weights.

			Grains.	Dwt.
24 grains	=	1 dwt.,		
20 dwt.	=	1 ounce	= 480	
12 ounces	=	1 pound	= 5760	= 240

Troy Weight is used for gold and silver. A cubic foot of pure gold is worth $362600, and a cubic inch $209.84.

A cubic foot of pure silver is worth $12338, and a cubic inch $7.14; gold being worth about 29.39 times that of silver.

7000 Troy grains	=	1 lb., Avoirdupois,
175 " pounds	=	144 lbs. "
175 " ounces	=	192 ounces, Avoirdupois,
437½ " grains	=	1 ounce "
1 " pound	=	.8228+ lb. "

Avoirdupois Weights.

27$\frac{11}{32}$ grains	=	1 dram,			
16 drams	=	1 ounce,	Drams.	Ounces.	Pounds.
16 ounces	=	1 pound =	256		
112 pounds	=	1 cwt. =	28672 =	1792	
20 cwt.	=	1 ton =	573440 =	35840 =	2240
28 pounds	=	1 quarter,			
4 quarters	=	1 cwt.,			
1 stone	=	14 lbs.			

The standard Avoirdupois pound is the weight of 27.7015 cubic inches of distilled water weighed in air, at the temperature of 39°.$\frac{84}{100}$, latitude of London; barometer, 30 inches.

Cubic or Solid Measures.

1728 cubic inches	=	1 cubic or solid foot,
27 cubic feet	=	1 cubic " yard,
128 cubic feet, 4×4×8	=	1 cord of wood,
24.75 " "	=	1 perch of stone,
306 " "	=	1 rod,
57.25 " "	=	1 chaldron = 36 bushels,
42 " "	=	1 ton of anthracite coal,
46 " "	=	1 " " bituminous "

Scripture Long Measures.

		Feet.	Inches.			Feet.	Inches.
A digit	=	0	0.912	A cubit	=	1	9.888
A palm	=	0	3.648	A fathom	=	7	3.552
A span	=	0	10.944				

Liquid Measures.

		Cubic Ins.
4 gills	= 1 pint	= 28.875
2 pints	= 1 quart =	57.750 = 8 gills,
4 quarts	= 1 gallon=	231. = 32 "
31½ gallons	= 1 barrel,	
42 gallons	= 1 tierce,	
2 bbls., or 63 gallons,	— 1 hogshead,	
2 tierces, or 84 gallons,	= 1 puncheon,	
2 hogsheads	= 1 pipe, or butt,	
2 pipes, or 252 gallons,	= 1 ton.	

· A standard gallon measures 231 cubic inches, and contains 8.3388822 lbs. Avoirdupois, or 58372.1754 Troy grains, of distilled water at 39° Fah., barometer at 30 inches.

Dry Measures.

			Pints.	Quarts.	Gallons.
2 pints	=	1 quart,			
4 quarts	=	1 gallon	= 8		
2 gallons	=	1 peck	= 16	= 8	
4 pecks	=	1 bushel	= 64	= 32	= 8
36 bushels	=	1 chaldron.			

The standard bushel is the Winchester, which contains 2150.42 cubic inches, or 77.627413 lbs. Avoirdupois of distilled water at its maximum density.

Its dimensions are 18½ inches diameter inside, 19½ inches outside, and 8 inches deep; and when heaped, the cone must not be less than 6 inches high, equal to 2747.715 cubic inches for a true cone.

A Gallon	=	268.8025 cubic inches.
A Struck Bushel	=	1.24445 cubic feet.
A Cubic Foot	=	.80356 of a Struck Bushel.
A Barrel of Flour	=	196 lbs.

Many articles, as seed, grain, etc., although measured by the bushel, are really sold by weight. The following are thus measured : Blue-grass seed, 14 lbs. to the bushel; Dried Apples, 33; Bran, 20; Oats, 35; Timothy, 45; Castor-beans, 46; Hemp-seed, 44; Barley, 48; Buckwheat, 52; Corn on the cob, 70; Salt, 85; Rye, Flaxseed, Corn, and Onions, 56; Potatoes, Beans, Wheat, and Clover-seed, 60, etc.

The Barrel is not a legal measure, excepting Flour, 196 lbs., and Beef, Pork, and Fish, 200 lbs. The capacity of a barrel varies, and is supposed to contain, Cement, 300 lbs.; Rice, 600; Powder, 25; Potatoes, $2\frac{1}{4}$ bushels, etc.

The Melting-Point of Metals and Effect on Bodies by Heat— Fahrenheit Thermometer.

	DEG.		DEG.
Cast-Iron	2754	Heat of common parlor-fire.	790
Tin	475	Boiling-point of mercury....	630
Lead	594	Boiling-point of linseed-oil.	600
Zinc	740	Boiling-point of alcohol	174
Brass	1900	Boiling-point of ether	98
Fine Silver	1850	Heat of human blood	98
Copper	2160	Vinous fermentation....60 to	77
Fine Gold	1983	Acetification begins at 78°	
Bismuth	487	and ends at	88
Tin and Bismuth, equal parts	283	Phosphorus burns at	43
Tin 3, Bis. 5, and Lead 2....	212	Snow and Salt, equal parts...	0
Platina	4593	Strong wines freeze at	—20
Antimony	955	Brandy freezes at	— 7
Gold, annealed	2266	Greatest cold produced	—90
Red heat visible in daylight	1077	Mercury melts	—39

Fluids boil in vacuo with 124° less heat than under the pressure of the atmosphere. All solids absorb heat when becoming fluid. The heat absorbed in liquefaction is given out again in freezing.

Water may be cooled at 20°.

Freezing water gives out 140° of heat, and when congealed to ice, the thickness of

9 *

2 inches will bear infantry,
4   "   "   " cavalry or light guns,
6   "   "   " heavy field-guns,
8   "   "   " 24-pounders on sledges, weight not
over 1000 lbs. to a square foot.

Table of the Temperature Required to Ignite Different Combustible Substances.

	DEG.		DEG.
Fulminating powder	374	Picrate powder, for torpedoes	570
Phosphorus	140	" " " muskets	576
Bisulphate of carbon vapor.	300	Charcoal, the most inflam-	
Fulminate of mercury	392	mable willow used for gun-	
Sulphur	400	powder	580
Equal parts of sulphur and		Charcoal, made by distilling	
chlorate of potash	395	wood at 500°	660
Gun-cotton	428	Charcoal made at 600°	700
Nitro-glycerine	494	Picrate powder, for cannon	716
Rifle-powder	550	Very dry pine wood	800
Gunpowder, coarse	563	" " oak "	900
Picrate of mercury, lead, or		Charcoal made at 800°	900
iron	565	" " " 1800°	1100
Aluminum	1832	" " " 2400°	1400

Weight of Nails.

NAME.	LENGTH.	NO. PER POUND.
	Ins.	
3-penny.	1	557
4 "	1¼	353
5 "	1¾	232
6 "	2	175
7 "	2¼	141
8 "	2½	101
10 "	2¾	68
12 "	3	54
20 "	3½	34

Weight of Metallic Balls.

DIAMETER IN INCHES.	CAST LEAD.	CAST COPPER.	CAST BRASS.	CAST IRON.	DIAMETER IN INCHES.	CAST LEAD.	CAST COPPER.	CAST BRASS.	CAST IRON.
	Lbs.	Lbs.	Lbs.	Lbs.		Lbs.	Lbs.	Lbs.	Lbs.
½	.026	.021	.019	.017	5¼	30.1	24.1	21.5	19.8
¾	.088	.070	.063	.058	5½	34.7	27.7	24.7	22.7
1	.209	.167	.148	.136	5¾	39.6	31.7	28.3	25.9
1¼	.408	.325	.290	.266	6	45.0	36.0	32.0	29.4
1½	.705	.562	.501	.460	6½	57.2	45.8	40.8	37.4
1¾	1.12	.893	.795	.731	7	71.5	57.2	50.9	46.8
2	1.67	1.33	1.19	1.07	7½	88.0	70.3	62.6	57.5
2¼	2.38	1.90	1.69	1.55	8	106.	85.3	76.0	69.3
2½	3.25	2.60	2.32	2.13	8½	127.	102.	91.2	83.7
2¾	4.34	3.47	3.09	2.83	9	151.	121.	108.	99.4
3	5.63	4.50	4.01	3.68	9½	178.	143.	127.	117.
3¼	7.15	5.72	5.10	4.68	10	208.	167.	148.	136.
3½	8.94	7.14	6.36	5.85	10½	241.	193.	172.	158.
3¾	11.0	8.79	7.83	7.19	11	277.	222.	198.	182.
4	13.4	10.7	9.50	8.73	11½	317.	253.	226.	207.
4¼	16.0	12.8	11.4	10.5	12	360.	288.	257.	236.
4½	18.9	15.2	13.5	12.4					
4¾	22.7	17.9	15.9	14.6	The weight of balls is as the				
5	26.0	20.8	18.6	17.0	cubes of their diameters.				

Copper.

To Ascertain the Weight of Copper.

RULE.—Find by calculation the number of cubic inches in the piece; multiply them by .32118, and the product will be the weight in pounds.

Lead.

To Ascertain the Weight of Lead.

RULE.—Find by calculation the number of cubic inches in the piece; multiply the sum by .41015, and the product will be the weight in pounds.

Brass.

To Ascertain the Weight of Ordinary Brass Castings.

RULE.—Find the number of cubic inches in the piece; multiply by .3112, and the product will be the weight in pounds.

Table Showing what Weight a Hemp Rope will Bear with Safety.

Circum-ference.	Pounds.	Circum-ference.	Pounds.	Circum-ference.	Pounds.
Ins.		Ins.		Ins.	
1	200	3½	2450	6	7200
1¼	312.5	3¾	2812.5	6¼	7812.5
1½	450	4	3200	6½	8450
1¾	612.5	4¼	3612.5	6¾	9112.5
2	800	4½	4050	7	9800
2¼	1012.5	4¾	4512.5	7¼	10512.5
2½	1250	5	5000	7½	11250
2¾	1512.5	5¼	5512.5	7¾	12012.5
3	1800	5½	6050	8	12800
3¼	2112.5	5¾	6612.5		

RULE.—Multiply the square of the circumference in inches by 200, and it gives the weight the rope will bear in pounds with safety. For the strength of a good hemp cable, multiply the square of the circumference by 120.

The specific gravity of a body is its weight as compared with that of water at a temperature of 60° Fah., and 30 inches at sea level.

At 60° pure water weighs 62.331 pounds avoirdupois per cubic foot.

To find the gravity of a body heavier than water, weigh it first in the air, then in water, and find the difference; the difference is what the body loses in water, and is the weight of a bulk of water equal to the bulk of the body. Then say as this difference: weight in air : : 1 : specific gravity of body.

Table of Specific Gravities.

Metals.	Average Specific Gravity.	Weight of a cub. ft. in lbs.
Aluminum	2.6	162
Antimony	6.70	418
Arsenic	5.76	360
Bismuth.	9.74	607
Brass	8.1	504
Bronze—copper 8 parts, tin 1, gun m. 8.5	8.5	529
Copper, cast	8.7	542
Gold, cast pure, 24 carat . . .	19.258	1204
Iron, cast	7.15	446
Iron, wrought	7.77	485
Lead	11.41	711
Mercury	13.62	849
Platinum	21.5	1342
Silver	10.5	655
Spelter, or Zinc	7.00	437.5
Steel	7.85	490
Tin, cast	7.35	459

Woods (Dry).

	Average Specific Gravity.	Weight of a cub. ft. in lbs.
Apple	.79	49
Ash	.75	47
Boxwood	.96	60
Cherry	.67	42
Chestnut	.66	41
Cork	.25	15.6
Ebony	1.33	83
Elm	.56	35
Hemlock	.40	25
Hickory	.85	53
Lignum-Vitæ	1.33	83
Logwood	.91	57

	Average Specific Gravity.	Weight of a cub. ft. in lbs.
Mahogany	.85	53
Maple	.79	49
Mulberry	.89	56
Oak	.95	59.3
Pine, White	.40	25
Pine, Yellow	.72	45
Poplar	.38	24
Spruce	.40	25
Sycamore	.59	37
Walnut	.61	38

Stones and Earths.

	Average Specific Gravity.	Weight of a cub. ft. in lbs.
Alabaster	2.7	168
Coal, Anthracite	1.5	93.5
Coal, Bituminous	1.35	84
Basalt	2.9	181
Borax	1.71	107
Brick, hard	......	150
Chalk	2.5	156
Clay, Potters', dry	1.9	119
Coke	......	23–32
Crystal, pure Quartz	2.66	165
Diamond	3.53	......
Earth, dry	......	80–92
Earth, soft	......	110
Emerald	2.7	......
Felspar	2.5	156
Flint	2.6	162
Garnet	4.2	......
Glass	2.98	186
Gneiss	2.69	168
Granite	2.69	168
Greenstone, trap	3	187

	Average Specific Gravity.	Weight of a cub. ft. in lbs.
Gypsum, Plaster of Paris	2.31	144
Hornblende	3.25	203
Limestone and Marbles	2.75	172
Mica	2.93	183
Peat, dry, compressed	......	20–30
Ruby and Sapphire	4.04	......
Salt	......	55
Salt, fine	......	49
Sand	2.65	106
Sandstone	2.41	150
Shales	2.6	162
Slate	2.8	175
Sulphur	2	125
Trap	3	187
Topaz	3.55	......

Gravel about the same as Sand.

Liquids.

Alcohol, common	.834	52.1
Alcohol, pure	.793	49.43
Ether	.716	44.6
Oil—Whale, Olive	.92	57.3
Petroleum	.878	54.8
Proof Spirit	.916	57.2
Turpentine	.87	54.3
Vinegar	1.080	68
Water, pure Rain or Distilled, at 32° Fahr., 30 in. Bar.	1.000	62.375
Water, Sea, average	1.028	64.08
Wine	.992	62
Wines, average	.998	62.3

Miscellaneous.

	Average Specific Gravity.	Weight of a cub. ft. in lbs.
Air, Atmospheric, at 60° Fahr., and under the pressure of one atmosphere, or 14.7 lbs. per square inch, weighs $\frac{1}{813}$ part as much as water at 60°	.00123	.0765
Asphaltum	1.4	87.3
Carbonic Acid Gas is $1\frac{1}{2}$ times as heavy as air	.00187	......
Cement	......	70
Charcoal, Pine and Oak	......	15–20
Fat	.93	58
Gutta-percha	.98	61.1
Hydrogen Gas is $14\frac{1}{2}$ times lighter than air, 16 times lighter than oxygen	......	.00527
Ice	.94	58.7
Ivory	1.82	114
Lard	.95	59.3
Lime	1.60	100
Naphtha	.848	52.9
Nitrate of Potash, or Saltpetre	200	......
Nitrogen Gas, $\frac{1}{35}$ part lighter than air	......	.0744
Oxygen Gas, $\frac{1}{10}$ part heavier than air	.00136	.0846
Pitch	1.15	71.7
Powder	1	62.3
Rosin	1.1	68.6
Snow, compacted by rain	......	50
Snow, fresh	......	5–12
Tar	1	62.4
Wax, Bees, average	.97	60.5

APPLICATION OF THE ABOVE.—When the weight of a body is required, find the contents of the body in cubic feet and multiply it by the factor in the table.

Gravitation.

TABLE EXHIBITING THE RELATION OF TIME, SPACE, AND VELOCITIES.

Seconds from beginning of descent	Velocity acquired at end of that time.	Squares.	Space fallen in that time.	Spaces.	Spaces fallen through in last second of fall.
	Ft.		Ft.		Ft.
1	32.166	1	16.08	1	16.08
2	64.333	4	64.33	3	48.25
3	96.5	9	144.75	5	80.41
4	128.665	16	257.33	7	112.58
5	160.832	25	402.08	9	144.75
6	193.	36	579.	11	176.91
7	225.166	49	788.08	13	209.08
8	257.333	64	1029.33	15	241.25
9	289.5	81	1302.75	17	273.42
10	321.666	100	1608.33	19	305.58
11	353.832	121	1946.08	21	337.75
12	336.	144	2316.	23	369.92

To find the velocity of a falling body, multiply the time in seconds by 32.166 for the velocity in feet per second.

Water.

Fresh water is 1 part oxygen and 2 parts hydrogen; by weight, 88.9 oxygen, 11.1 hydrogen. One cubic inch at 62°, barometer 30 inches, weighs 252.458 grains; from this it expands either from cold or heat. The temperature of 32° reduces it to solid ice, which expansion is about $\frac{1}{12}$ part of its original bulk as water, and this expansive force is sufficient to split iron water-pipes; by this expansive force rocks are split, and walls that are not of sufficient depth in the earth, are lifted upwards and overthrown.

Wood remains sound for centuries, under either fresh or salt water, if not exposed to the action of the air or strong currents. Hard water contains considerable lime; water in lead pipes produces carbonate of lead, an active poison; but

10

where lime is present in the water, it forms a coating inside, that prevents the poisonous formation. Many substances are held in solution, being extracted from the earth through which the water penetrates.

Water is the sole product of the combustion of hydrogen in oxygen, or the atmosphere; according to Humboldt, 8 parts by weight of O. and 1 part of H. It is the important compound of all chemical mixtures. It comprises the greatest part of the earth's surface in the form of oceans, seas, lakes, and rivers; and in the north and south poles, snow, ice, iceberg, and glacier rising in form of vapor to the atmosphere. It produces by condensation mist, rain, and snow, to be returned to their original condition, to go through the same process constantly until the end of time.

In the vegetable kingdom it is ever present, varying in proportion from 10 to 90 per cent.; dry wood contains 20 per cent. The human body, weighing 150 pounds, contains 110 to 120 lbs. of water; the rocks contain it in more or less quantities, according to their construction; gypsum, 20 per cent.

Rain-water collected in cities or near manufacturing districts is never pure, because it partakes of and contains gases which are given off from, or developed by, the combustion of coal, etc., forming compounds of sulphur and other substances.

After thunder-storms the rain-water is always found to contain minute quantities of nitric acid, produced from the component parts of the air, nitrogen and oxygen, combining from the action of the lightning.

Rain-water almost always contains a little organic matter, and it will become putrid after standing some time; organic substances are taken up into the clouds sometimes in large bodies, and rained down, causing astonishment, wonder, and superstition.

Water is the product of the combustion of hydrogen in oxygen, or the atmosphere, by action of lightning or electricity. The action of the voltaic current also decomposes water, setting free its component gases, the combustible hydrogen, and the supporter of combustion, oxygen. And may we not anticipate that in the not far distant future that water will be used as our combustible material. When that takes place, water in the form of steam will cease to be the power, and water itself will be the substitute.

Acidulous Springs.— Waters that are charged with carbonic acid in such quantities as to cause them to sparkle and effervesce when flowing from the springs are called acidulous. On account of the solvent power of this acid upon limestone and other rocks, such waters hold in solution lime, magnesia, and iron. When the latter is present, one grain or more to the gallon, the spring is a chalybeate.

Almost all spring waters contain minute quantities of iron, generally in the form of bicarbonate. The Saratoga waters contain from two to three grains of this compound per gallon.

Water for drinking should be boiled, then left to become cool and aerated.

Distilled water must be thoroughly aerated to render it palatable and wholesome.

Charcoal has the property of purifying water contaminated by organic matter. When rain-water cisterns become foul, it is a common practice to throw in a bushel or two of fresh charcoal. Permanganate of potassa is also used by travellers, who carry with them a small vial of the crystallized salt: a small particle added to a glass of water renders it pure in a few moments. Impure water placed in casks on shipboard sometimes undergoes a kind of fermentation, by which the impurities are worked off and the water rendered wholesome.

A calculation has been made by which it appears that 36 cubic miles of water are poured into the ocean daily by the rivers, and it would take 30,000 years for all the water to rise as vapor and fall as rain, and make one trip back to the ocean.

Sea-water, according to the analysis of Dr. Murray, at the specific gravity of 1.039, contains

Muriate of soda,	220.01	=	$\frac{1}{46}$
Sulphate of soda,	33.16	=	$\frac{1}{302}$
Muriate of magnesia,	42.08	=	$\frac{1}{238}$
Muriate of lime,	7.84	=	$\frac{1}{1276}$
	303.09	=	$\frac{1}{33}$

Of salt water, a cubic foot weighs 64.3 lbs.; a cubic inch, .3721 lbs.

Saline Contents of Sea-Water from Different Localities.

Baltic	. . .	6.60	Equator .	. .	39.42
Black Sea	. .	21.60	South Atlantic	.	41.20
Arctic	. . .	28.30	Sea of Marmora	.	42.00
Irish Sea .	. .	33.76	North Atlantic	.	42.60
British Channel	.	35.50	Dead Sea	.	385.00
Mediterranean .	.	39.40			

There are 62 volumes of carbonic acid in 1000 parts of sea-water, and all the metallic substances are held in solution in small proportions.

Analysis of One Gallon of Sea-Water from the Atlantic Ocean, made by Von Bilbra.

SPECIFIC GRAVITY.......... 1.0275

Chloride of Sodium	1671.34	Chloride of Iron	.	Trace
Chloride of Magnesium .	. . 199.66	Bromide of Sodium.		31.16
		Iodide of Sodium	.	Trace

Sulphate of Potassa	108.46	Silver . . .	Trace
Sulphate of Magne-		Copper . .	"
sia . . .	34.99	Lead . . .	"
Sulphate of Lime .	93.30	Arsenic . .	"
Phosphate of Soda	Trace	Silica . . .	"
Carbonate of Lime	"	Organic matter .	"

Total grains in U. S. gallon . . 2138.91

Percentage by weight 3.569
Water 96.431
                                    ———
                                    100
Weight of one gallon is 59.922 grains.

Dead Sea water, percentage by weight . . . 19.733
Water 80.267
                                    ———
                                    100
Weight of one gallon is 68.352 grains.

The Herepaths' Analysis.

	Density.	Grains Saline Matter in One Gallon.	Oz. Saline Matter in One Gallon.
The Atlantic Ocean................	1.027	2.139	4.89
The Dead Sea....................	1.172	13.488	30.86
The Great Salt Lake.............	1.170	15.203	34.72
Lake Ooroomeeyah, in Persia..	1.188	18.209	41.69

Wind.

Atmospheric air extends about 45 miles from the surface of the earth. Its component parts are $\frac{4}{5}$ nitrogen and $\frac{1}{5}$ oxygen, or 77, N., 23, O. It generally contains a trace of carbon, hydrogen, and ammonia. Greatest known heat of the air in the sun, 145° Fah.; greatest known cold below zero, 65° Fah.

The mean pressure generally admitted is 14.7 lbs. per

10 *   H

square inch; barom. 30 and 34 feet water; specific gravity compared with water, .0012046.

The mean weight of a column of air a foot square, and of the altitude of the atmosphere, is equal to 2116.8 lbs. avoirdupois; the rate of expansion, and also all elastic fluids for all temperatures, is uniform; from 32° to 212° they expand from 1000 to 1376, equal to $\frac{1}{479}$* of their bulk for every degree of heat.

At 7 miles from the earth's surface, the air is 4 times rarer or lighter than at the earth's surface; at 14 miles, 16 times; at 21 miles, 64 times, and so continues in the same ratio. At a temperature of 33°, the mean velocity of sound is 1100 feet per second, and is increased or diminished half a foot for each degree above or below 33°.

* $\frac{1}{479}$ equals .002087 for each degree.

Velocity and Pressure of Wind.

Miles per hour.	Feet per minute.	Pressure per square foot in avoir. lbs.	REMARKS.
1	88	.005	Barely observable.
2	176	.020	} Perceptible.
3	264	.045	
4	352	.080	Pleasant breeze.
5	440	.125	}
6	528	.180	} Gentle, pleasant wind.
8	704	.320	}
10	880	.500	} Brisk blow.
15	1320	1.125	
20	1760	2.000	} Very brisk.
25	2200	3.125	
30	2640	4.500	} High wind.
35	3080	6.125	
40	3520	8.000	} Very high.
45	3960	10.125	
50	4400	12.500	Storm.
60	5280	18.000	Violent storm.
80	7040	32.000	Hurricane.
100	8800	50.000	Tornado.

Traction.

The tractive power of a horse diminishes as his speed increases. The traction of a horse on a level road, pulling for ten hours in the day, is as follows:

Miles per hour.	Lbs. Traction.	Miles per hour.	Lbs. Traction.
¾	333.33	2¼	111.11
1	250.	2½	100.
1¼	200.	2¾	90.91
1½	166.66	3	83.33
1¾	142.86	3½	71.43
2	125.	4	62.50

A horse travels 400 yards, at a walk, in 4½ minutes; at a trot, in 2 minutes; at a gallop, 1 minute. Average weight, 1,000 lbs. Carrying a soldier and equipments, 225 lbs., travels 25 miles a day of 8 hours.

A draft horse will draw 1,600 lbs. 23 miles in one day, wagon included. Generally, the work allowed is equal to 22,500 lbs., raised 1 foot in a minute, for 8 hours a day.

A man of ordinary strength exerts a force of 30 lbs. for 10 hours in a day, with a velocity of 2¼ feet in a second = 4500 lbs. raised 1 foot in a minute = ⅕ the work of a horse.

A foot soldier travels in one minute, in common time, 90 steps = 70 yards; in quick time, 110 steps = 86 yards; in double-quick time, 140 steps = 109 yards, and occupies in the ranks a front of 20 inches and a depth of 13, without a knapsack. Interval between the ranks, 13 inches; average weight of man, 150 lbs. Five men can stand in a space of 1 yard square.

A man travels without a load, on level ground, during 8½ hours a day, at the rate of 3.7 miles per hour, or 31¼ miles per day. He can carry 111 lbs. 11 miles per day.

Compressibility of Liquids.

BY L. CAILLETET.

	Density (sp. gr.).	Temperature.	Compressibility.	No. of atmospheric pres're.
Distilled water, free from air	1.000	8°	0.0000451	705
Sulphide of carbon..............	......	8°	0.0000980	607
		9°	0.0000676	174
Alcohol.............................	0.858	9°	0.0000701	305
		11°	0.0000727	680
Petroleum oil.....................	0.865	11°	0.0000828	610
Petroleum essence, benzoline	0.720	10°.5	0.0000981	630
Sulphuric ether...................	......	10°	0.0001440	630
Sulphuric acid (fluid)..........	......	14°	0.0003014	606

Compressibility.

BY MR. CANTON.

Spirit of Wine.	.	.	.	.	0.000066 of its bulk.
Olive Oil.	.	.	.	.	0.000048 " "
Rain-Water	.	.	.	.	0.000046 " "
Sea "	.	.	.	.	0.000046 " "
Mercury .	.	.	.	.	0.000003 " "

To Measure Round Timber.

Multiply the length in inches by the square of ¼ the mean girth in inches, and the product divided by 1728 will give the contents in cubic feet.

When the length is given in feet and the girth in inches, divide by 144.

When all the dimensions are in feet, the product is the contents without a divisor.

EXAMPLE.—The girth of a piece of timber 31.416 and

62.832 inches, and its length 50 feet: required its contents.
$\frac{31.416 + 62.832}{2} \div 4 = 11.781$, and $11.781^2 \times 50 \div 144 =$
48.1916 cubic feet.

Capacity of Cisterns in U. S. Gallons for each 10 Inches in Depth.

2	feet in	diameter	19.5	8	feet in	diameter	313.33
2½	"	"	30.6	8½	"	"	353.72
3	"	"	44.06	9	"	"	396.56
3½	"	"	59.97	9½	"	"	461.40
4	"	"	78.33	10	"	"	489.20
4½	"	"	99.14	11	"	"	592.40
5	"	"	122.40	12	"	"	705.
5½	"	"	148.10	13	"	"	827.4
6	"	"	176.25	14	"	"	959.6
6½	"	"	206.85	15	"	"	1101.6
7	"	"	239.88	20	"	"	1958.4
7½	"	"	275.40	25	"	"	3059.9

Hills in an Acre of Ground.

40	feet apart,	27	hills.	8	feet apart,	680	hills.
35	"	35	"	6	"	1210	"
30	"	48	"	5	"	1742	"
25	"	69	"	3½	"	3556	"
20	"	108	"	3	"	4840	"
15	"	193	"	2½	"	6969	"
12	"	302	"	2	"	10980	"
10	"	435	"	1	"	43560	"

Alloys.

COMPOSITIONS.	COPPER.	ZINC.	TIN.	NICKEL.	LEAD.	ANTI-MONY.	BISMUTH.	SILVER.	COBALT OF IRON.
Chinese white copper	40.4	25.4	2.6	31.6					
Chinese silver............	65.2	19.5	......	13.	......	...	...	2.	12
White argentane......	8.	3.5	......	3.					
Pinchbeck..............	5.	1.							
German silver.........	1.	1.	......	1.					
Britannia metal......	......	......	4.	......	......	4.			
When fused, add...	......	......	......	......	.....	4.	4.		
Printing metal.........	......	......	......	......	4.	1.			
Small type and ster- } cotype plates. }	......	......	......	......	9.	2.	2.		
Telescopic mirrors....	100.	......	50.						
Bronze statuary........	91.4	5.5	1.4	......	1.7				
Large cannon..........	90.	......	10.						
Small cannon...........	93.	......	7.						
Medals..................	100.	......	8.						
Cymbals................	80.	......	20.						
Tutenag copper........	8.	5.	......	3.					
Newton's fusible } metal. It melts at a temperature } less than that of boiling water. }	......	......	3.	......	5.	...	8.		
A metal that ex- } pands in cooling. }	......	......	......	......	9.	4.	1.		

The more infusible metals should be melted first.

----

23 cubic feet of sand, or 18 cubic feet of earth, or 17 cubic feet of clay, make a ton.

18 cubic feet of gravel or earth before digging make 27 cubic feet when dug.

A chaldron of bituminous coal yields about 10.000 cubic feet of gas, and 1.43 cubic feet of gas per hour give a light equal to one good candle. 3 cubic feet = 10 candles.

10 cubic yards of meadow hay weigh a ton. When the hay is taken out of large or old stacks, 8 and 9 yards will make a ton.

Cast-iron expands $\frac{1}{162000}$ of its length for one degree of heat; greatest change in the shade in this climate, $\frac{1}{1170}$ of its length; exposed to the sun's rays, $\frac{1}{1000}$; shrinks in cooling from $\frac{1}{85}$ to $\frac{1}{95}$ of its length; will bear, without permanent alteration, 15.300 lbs. upon a square inch, and an extension of $\frac{1}{1200}$ of its length.

Wrought iron expands $\frac{1}{143000}$ of its length for one degree of heat; will bear on a square inch, without permanent alteration, 17.800 lbs., and an extension in length of $\frac{1}{1100}$; cohesive force is diminished $\frac{1}{3000}$ by an increase of 1 degree of heat compared with cast-iron; its strength is 1.12 times, its extensibility 0.86 times, and its stiffness 1.3 times.

Comparative Weight of Timber in a Green and a Seasoned State.

TIMBER.	WEIGHT OF CUBIC FOOT.	
	Green.	Seasoned.
	Lbs. Oz.	Lbs. Oz.
English Oak	71.10	43.8
Cedar	32.	28.4
Riga Fir	48.12	25.8
American Fir	44.12	30.11
Elm	66.8	37.5
Beech	60.	53.6
Ash	58.3	50.

The average weight of the timber materials in an English vessel of war is about 50 lbs. to the cubic foot, and for masts and yards about 40 lbs.

Table of Board Measure.

THE first left-hand column contains the breadth in inches. The top of the columns contains the length in feet.

WIDTH	7 FT.	8 FT.	9 FT.	10 FT.	11 FT.	12 FT.	13 FT.	14 FT.	15 FT.	16 FT.	17 FT.	18 FT.
3	1.9	2.0	2.3	2.6	2.9	3	3.3	3.6	3.9	4.0	4.3	4.6
4	2.4	2.8	3.0	3.4	3.8	4	4.4	4.8	5.0	5.4	5.8	6.0
5	2.11	3.4	3.9	4.2	4.7	5	5.5	5.10	6.3	6.8	7.1	7.6
6	3.6	4.0	4.6	5.0	5.6	6	6.6	7.0	7.6	8.0	8.6	9.0
7	4.1	4.8	5.3	5.10	6.5	7	7.7	8.2	8.9	9.4	9.11	10.6
8	4.8	5.4	6.0	6.8	7.4	8	8.8	9.4	10.0	10.8	11.4	12.0
9	5.3	6.0	6.9	7.6	8.3	9	9.9	10.6	11.3	12.0	12.9	13.6
10	5.10	6.8	7.6	8.4	9.2	10	10.10	11.8	12.6	13.4	14.2	15.0
11	6.5	7.4	8.3	9.2	10.1	11	11.11	12.10	13.9	14.8	15.7	16.6
12	7.0	8.0	9.0	10.0	11.0	12	13.0	14.0	15.0	16.0	17.0	18.0
13	7.7	8.8	9.9	10.10	11.11	13	14.1	15.2	16.3	17.4	18.5	19.6
14	8.2	9.4	10.6	11.8	12.10	14	15.2	16.4	17.6	18.8	19.10	21.0
15	8.9	10.0	11.3	12.6	13.9	15	16.3	17.6	18.9	20.0	21.3	22.6
16	9.4	10.8	12.0	13.4	14.8	16	17.4	18.8	20.0	21.4	22.8	24.0
17	9.11	11.4	12.9	14.2	15.7	17	18.5	19.10	21.3	22.8	24.1	25.6
18	10.6	12.0	13.6	15.0	16.6	18	19.6	21.0	22.6	24.0	25.6	27.0
19	11.1	12.8	14.3	15.10	17.5	19	20.7	22.2	23.9	25.4	26.11	28.6
20	11.8	13.4	15.0	16.8	18.4	20	21.8	23.4	25.0	26.8	28.4	30.0
21	12.3	14.0	15.9	17.6	19.3	21	22.9	24.6	26.3	28.0	29.9	31.6
22	12.10	14.8	16.6	18.4	20.2	22	23.10	25.8	27.6	29.4	31.2	33.0
23	13.5	15.4	17.3	19.2	21.1	23	24.11	26.10	28.9	30.8	32.7	34.6
24	14.0	16.0	18.0	20.0	22.0	24	26.0	28.0	30.0	32.0	34.0	36.0
25	14.7	16.8	18.9	20.10	22.11	25	27.1	29.2	31.3	33.4	35.5	37.6
26	15.2	17.4	19.6	21.8	23.10	26	28.2	30.4	32.6	34.8	36.10	39.0
27	15.9	18.0	20.3	22.6	24.9	27	29.3	31.6	33.9	36.0	38.3	40.6
28	16.4	18.8	21.0	23.4	25.8	28	30.4	32.8	35.0	37.4	39.8	42.0
29	16.11	19.4	21.9	24.2	26.7	29	31.5	33.10	36.3	38.8	41.1	43.6
30	17.6	20.0	22.6	25.0	27.6	30	32.6	35.0	37.6	40.0	42.6	45.0

Table of Board Measure—*Continued.*

WIDTH	19 FT.	20 FT.	21 FT.	22 FT.	23 FT.	24 FT.	25 FT.	26 FT.	27 FT.	28 FT.	29 FT.	30 FT.
3	4.9	5.0	5.3	5.6	5.9	6	6.3	6.6	6.9	7.0	7.3	7.6
4	6.4	6.8	7.0	7.4	7.8	8	8.4	8.8	9.0	9.4	9.8	10.0
5	7.11	8.4	8.9	9.2	9.7	10	10.5	10.10	11.3	11.8	12.1	12.6
6	9.6	10.0	10.6	11.0	11.6	12	12.6	13.0	13.6	14.0	14.6	15.0
7	11.1	11.8	12.3	12.10	13.5	14	14.7	15.2	15.9	16.4	16.11	17.6
8	12.8	13.4	14.0	14.8	15.4	16	16.8	17.4	18.0	18.8	19.4	20.0
9	14.3	15.0	15.9	16.6	17.3	18	18.9	19.6	20.3	21.0	21.9	22.6
10	15.10	16.8	17.6	18.4	19.2	20	20.10	21.8	22.6	23.4	24.2	25.0
11	17.5	18.4	19.3	20.2	21.1	22	22.11	23.10	24.9	25.8	26.7	27.6
12	19.0	20.0	21.0	22.0	23.0	24	25.0	26.0	27.0	28.0	29.0	30.0
13	20.7	21.8	22.9	23.10	24.11	26	27.1	28.2	29.3	30.4	31.5	32.6
14	22.2	23.4	24.6	25.8	26.10	28	29.2	30.4	31.6	32.8	33.10	35.0
15	23.9	25.0	26.3	27.6	28.9	30	31.3	32.6	33.9	35.0	36.3	37.6
16	25.4	26.8	28.0	29.4	30.8	32	33.4	34.8	36.0	37.4	38.8	40.0
17	26.11	28.4	29.9	31.2	32.7	34	35.5	36.10	38.3	39.8	41.1	42.6
18	28.6	30.0	31.6	33.0	34.6	36	37.6	39.0	40.6	42.0	43.6	45.0
19	30.1	31.8	33.3	34.10	36.5	38	39.7	41.2	42.9	44.4	45.11	47.6
20	31.8	33.4	35.0	36.8	38.4	40	41.8	43.4	45.0	46.8	48.4	50.0
21	33.2	35.0	36.9	38.6	40.3	42	43.9	45.6	47.3	49.0	50.9	52.6
22	34.10	36.8	38.6	40.4	42.2	44	45.10	47.8	49.6	51.4	53.2	55.0
23	36.5	38.4	40.3	42.2	44.1	46	47.11	49.10	51 9	53.8	55.7	57.6
24	38.0	40.0	42.0	44.0	46.0	48	50.0	52.0	54.0	56.0	58.0	60.0
25	39.7	41.8	43.9	45.10	47.11	50	52.1	54.2	56.3	58.4	60.5	62.6
26	41.2	43.4	45.6	47.8	49.10	52	54.2	56.4	58.6	60.8	62.10	65.0
27	42.9	45.0	47.3	49.6	51.9	54	56.3	58.6	60.9	63.0	65.3	67.6
28	44.4	46.8	49.0	51.4	53.8	56	58.4	60.8	63.0	65.4	67.8	70.0
29	45.11	48.4	50.9	53.2	55.7	58	60.5	62.10	65.3	67.8	70.1	72.6
30	47.6	50.0	52.6	55.0	57.6	60	62.6	65.0	67.6	70.0	72.6	75.0

EXAMPLE.—To find the contents of a board 19 feet long and 25 inches broad, look for the column headed 19 feet, and in the first column for 25 inches; in the line of meeting of the breadth and length, you have 39 feet, 7 inches, as the contents.

11

Morse Telegraph Alphabet.

Abbreviations.		Abbreviations.
G. M.—Good-morning.	A B C D E F G H I	Ahr.—Another.
G. N.—Good-night.	J K L M N O P	Ans.—Answer.
Immy.—Immediately.	Q R S T U V W	Bk.—Back.
Impt.—Important.	X Y Z	Bf.—Before.
Min.—Minute.		Bn.—Been.
Msg.—Messenger.		Bat.—Battery.
Msk.—Mistake.	Period. Comma. Exclamation. Interrogation. Paragraph. Parenthesis.	Bbl.—Barrel.
No.—Number.		Col.—Collect.
Ngt.—Nothing.	1 2 3 4 5 6	Ck.—Check.
N. M.—No more.	7 8 9 0	Co.—Company.
O. K.—Correct.		D. H.—Free.
Ofs.—Office.		E. X.—Express.
Op.—Operator.		Frt.—Freight.
P. D.—Paid.		G. A.—Go ahead.
Q. K.—Quick.		P. O.—Post-Office.

AND ABBREVIATIONS.

The Signals Mostly Used are as follows:

1.—Wait a minute.	13.—What is the matter?
4.—Shall I go ahead?	77.—I have a message for you.
5.—Have you anything for me?	41.—Answer quick by telegraph.
13.—Do you understand?	73.—Accept my compliments.
134.—Who are you at the key?	

Resistance Measurement of No. 9 Galvanized Iron Wire.

OHMS.	MILES.	OHMS.	MILES.	OHMS.	MILES.	OHMS.	MILES.	OHMS.	MILES.
$\frac{1}{10}$	41 f	106^{6}	8	230	17¼	390	29¼	675	50⅝
$\frac{5}{10}$	204 f	110	8¼	235	17⅝	400	30	680	51
1	406 f	115	8⅝	240	18	410	30¾	700	52½
2	813 f	120	9	245	18⅜	413^{3}	31	720	54
3	¼	125	9⅜	250	18¾	420	31½	725	54⅜
5	⅜	130	9¾	253^{3}	19	426^{6}	32	750	56¼
10	$\frac{6}{8}$	133^{3}	10	255	19⅛	430	32¼	760	57
13^{3}	1	135	10⅜	260	19½	440	33	775	58⅛
15	1⅛	140	10½	265	19⅞	450	33¾	800	60
20	1½	145	10¾	266^{6}	20	453^{3}	34	825	61⅞
25	1⅞	146^{6}	11	270	20¼	460	34½	840	63
26^{6}	2	150	11¼	275	20⅝	466^{6}	35	850	63¾
30	2¼	155	11⅝	280	21	470	35¼	875	65⅝
35	2⅝	160	12	285	21⅜	480	36	880	66
40	3	165	12⅜	290	21¾	490	36¾	900	67½
45	3⅜	170	12¾	293^{3}	22	493^{3}	37	920	69
50	3¾	173^{3}	13	295	22⅛	500	37½	925	69⅜
53^{3}	4	175	13⅛	300	22½	506^{6}	38	950	71¼
55	4⅛	180	13½	306^{6}	23	510	38¼	960	72
60	4½	185	13⅞	310	23¼	520	39	975	73⅛
65	4⅞	186^{6}	14	320	24	530	39¾	1,000	75
66^{6}	5	190	14¼	330	24¾	533^{3}	40	2,000	150
70	5¼	195	14⅝	333^{3}	25	540	40½	3,000	225
75	5⅝	200	15	340	25½	546^{6}	41	4,000	300
80	6	205	15⅜	346^{6}	26	550	41¼	5,000	375
85	6⅜	210	15¾	350	26¼	560	42	6,000	450
90	6¾	213^{3}	16	360	27	575	43⅛	7,000	525
93^{3}	7	215	16⅛	370	27¾	600	45	8,000	600
95	7⅛	220	16½	373^{3}	28	625	46⅜	9,000	675
100	7½	225	16⅞	380	28½	640	48	10,000	750
105	7⅞	226^{6}	17	386^{6}	29	650	48¾	20,000	1500

Table of Number of Feet Bare Copper Wire to the Pound.

Birmingham Wire Gauge.	Diameter in Inches.	Number of Feet Bare Wire in Lb.	Birmingham Wire Gauge.	Diameter in Inches.	Number of Feet Bare Wire in Lb.
1	.300	3.52	21	.035	280.1
2	.280	4.44	*22	.030	381.6
3	.260	5.09	*23	.027	450.0
4	.240	6.00	*24	.025	533.1
5	.220	7.11	25	.023	630.0
6	.200	8.61	*26	.019	923.4
7	.185	10.09	27	.018	1029.0
8	.170	11.91	*28	.016	1302.0
9	.155	14.32	29	.015	1481.4
10	.140	17.50	*30	.014	1708.5
11	.125	22.1	31	.012	2314.8
*12	.110	28.4	*32	.010	3333.3
*13	.095	37.2	33	.009	3421.4
*14	.085	48.0	*34	.0096	3616.8
15	.075	61.3	35	.0087	4399.8
*16	.065	81.9	36	.0079	5340.0
17	.057	106.1	37	.0067	7425.6
*18	.050	137.9	38	.0058	9908.7
19	.045	161.3	39	.0042	18996.1
*20	.040	215.6	40	.0039	21915.0

*Standard sizes.

Weight of Insulated Office Wires.

DESCRIPTION.	SIZE STUBB'S GAUGE.	FEET PER LB.
Kerite..	14	33
"	16	50
Gutta-Percha................................	14	55
" "	16	65
Braided, paraffined, and compressed....	14	48
" " " "	16	60
" " " "	18	75
Paraffined, single cover....................	19	175
Cotton covered, plain........	16	85
" " "	18	115
" " "	19	180
" " "	20	215

Table of Iron Wire.

Birming- ham Wire Gauge.	Diameter in Inches.	Weight 100 feet.	Weight One Mile. Galvan- ized.	Weight One Mile, not Gal- vanized.	Breaking Strains.	Length of Bundles in Feet.
0 ...	0.340	29.44	1490	1416	7280	213
1 ...	0.300	22.92	1210	1150	5650	273
2 ...	0.280	19.97	1054	1002	4930	315
3 ...	0.260	17.22	909	854	4250	363
4 ...	0.240	11.00	775	747	3620	429
5 ...	0.220	12.34	651	619	3040	510
6 ...	0.200	10.19	538	512	2510	609
7 ...	0.185	8.72	461	438	2220	717
8* ...	0.170	7.37	389	370	1840	853
9* ...	0.155	6.12	323	307	1560	1026
10* ...	0.140	4.99	264	251	1280	1260
11* ...	0.125	3.98	211	200	1000	1587
12* ...	0.110	3.08	163	157	800	2100
13 ...	0.095	2.35	124	118	568	2679
14 ...	0.085	1.84	97	93	456	3426
15 ...	0.075	1.43	76	73	352	4404
16 ...	0.065	1.08	57	55	264	5862
17 ...	0.057	0.83	44	42	208	7620
18 ...	0.050	0.64	34	32	160	9450
19 ...	0.045	0.52	27	26	128	12255
20 ...	0.040	0.41	21	20	104	14736

*Those marked with a star are standard sizes for telegraph use.

The Mail Steamer *Baltic* ran from Liverpool to New York, in 1853, in 9 days, 16 hours, and 33 minutes; and from New York to Aspinwall, in 1859, in 6 days and 16 hours, and returned in 6 days and 22 hours — distance, 1980 miles each way.

The Mail Steamer *Adriatic* ran from New York to Liverpool, in 1860, in 9 days, 13 hours, and 30 minutes; from Galway to Quarantine, New York, in 8 days, 12 hours, and 30 minutes, touching at St. John's, N. B. On a passage from New York to Liverpool she ran 365 knots in 24 hours, and in 1861 she ran a measured mile in England in 3 minutes and 18 seconds = 18.181 knots per hour.

11 *

The Mail Steamer *Vanderbilt*, of New York, ran from New York to the Needles, in 1857, in 9 days and 8 hours; and from the Needles to New York, in 1859, in 9 days, 9 hours, and 26 minutes.

British and North American Royal Mail Steamship Company's Steamer *Persia*, of Glasgow, ran from New York to Liverpool, in 1856, in 9 days, 1 hour, and 45 minutes; and from Liverpool to New York, in 1861, in 9 days, 18 hours, and 1 minute.

Steamer *Ocean Bird*, of New York, ran from New York to Havana in 4 days and 4 hours.

Steamship *Daniel Drew*, of New York, ran from Jay Street, New York, to Albany, in 7 hours and 20 minutes; tide favorable, but wind ahead. Her time to Hudson, 125 miles, deducting landings, was 5 hours and 5 minutes; speed through the water fully 22.3 miles per hour.

Sovereign of the Seas, (clipper ship,) of Boston, in 22 days sailed 5391 knots, or 245 knots per day. For 4 days she sailed 341.78 knots per day, and for 1 day 375 knots.

Northern Light, (clipper ship,) of Boston, sailed from San Francisco to Boston in 76 days and 8 hours.

Flying Cloud, (clipper ship,) of Boston, sailed 374 knots in 1 day.

Red Jacket, (clipper ship,) of New York, sailed from New York to Liverpool in 13 days, 1 hour, and 25 minutes.

Nightingale, (clipper ship,) of New York, sailed from New York to Melbourne in 73 days.

Decimal Table

FOR REDUCING VULGAR TO DECIMAL FRACTIONS.—In a line with the numerator and under the denominator is the equivalent decimal fraction; thus, $\frac{9}{12} = 0.75$; $\frac{6}{16} = 0.375$. 8ths and 4ths may be had from 16ths; 6ths and 3ds may be had from 12ths.

NUMERATOR.	DENOMINATOR.	DENOMINATOR.	DENOMINATOR.
	12ths.	16ths.	96ths.
1	0.08333	0.0625	0.0104166
2	0.16666	0.1250	0.0208333
3	0.25000	0.1875	0.0312500
4	0.33333	0.2500	0.0416666
5	0.41666	0.3125	0.0520833
6	0.50000	0.3750	0.0625000
7	0.58333	0.4375	0.0729166
8	0.66666	0.5000	0.0833333
9	0.75000	0.5625	Eighths of an
10	0.83333	0.6250	inch reduced
11	0.91666	0.6875	to the deci-
12	1.00000	0.7500	mal of a foot.
13	....	0.8125	
14	....	0.8750	
15	....	0.9375	

Reduce 6 feet $4\frac{1}{2}$ inches to the decimal of a foot by the above.

$$6.00000 = 6 \text{ feet.}$$
$$.33333 = 4 \text{ inches.}$$
$$.04167 = 4 \text{ 8ths.}$$

Feet, $\overline{6.37500}$

Heat-Radiating Power of Different Bodies.

Water	100	Bright lead	19	Ice	85
Lamp-black	100	Silver	12	Mercury	20
Writing-paper	100	Blackened tin	100	Polished iron	15
Glass	90	Clean tin	12	Copper	12
India-ink	88	Scraped tin	16		

Table for Finding the Distance of Objects at Sea in Statute Miles.

Height in Feet.	Distance in Miles.	Height in Feet.	Distance in Miles.	Height in Feet.	Distance in Miles.	Height in Feet.	Distance in Miles.
*.582	1.	11	4.39	30	7.25	200	18.72
1	1.31	12	4.58	35	7.83	300	22.91
2	1.87	13	4.77	40	8.37	400	26.46
3	2.29	14	4.95	45	8.87	500	29.58
4	2.63	15	5.12	50	9.35	1000	32.41
5	2.96	16	5.29	60	10.25	2000	59.20
6	3.24	17	5.45	70	11.07	3000	72.50
7	3.49	18	5.61	80	11.83	4000	83.7
8	3.73	19	5.77	90	12.55	5000	93.5
9	3.96	20	5.92	100	13.23	1 mile.	96.1
10	4.18	25	6.61	150	16.20		

* 6.99 inches.

The difference in two levels is as the square of the distance. Thus, if the height is required for 2 miles:

$$1^2 : 2^2 :: 6.99 : 27.96 \text{ inches.}$$

For geographical miles, the distance for one mile is 7.962 inches.

EXAMPLE.—If a man at the fore-top-gallant masthead of a ship, 100 feet from the water, sees another and a large ship (hull to), how far are the ships apart?

A large ship's bulwarks are, say 20 feet from the water.

Then, by table, 100 feet　　=　　13.23
20 "　　=　　5.92
　　　　　　　　　　　　　　　　　　————

Distance, 19.15 miles.

NOTE.—$\frac{1}{13}$ should be added for horizontal refraction.

Heat-Conducting Power of Different Bodies.

Gold	1000	Marble	24	Zinc	363
Silver	973	Fire-brick	11	Lead	180
Iron	374	Platinum	381	Porcelain	12.2
Tin	304	Copper	898	Fire-clay	11.4

Products Obtained from Coal.

BY HENRY A. MOTT, JR.

COAL.

Gas, illuminating, etc.
Tar.........
Ammonia
Water.
Coke, for Fuel.

Oils, 30 %

Naphtha
Benzole { Benzole } { Used to make
{ Toluol } { Aniline.
Naphtha . . Used for Varnishes.
Xylole . . . Used for Small-pox.

FURNISHES

Dead Oil
Carbolic Acid } { Used for Disin-
Cresylic Acid } { fectants.
Naphthaline . . . Dyes, etc.
Anthracene ½ % } { Used to make
Chrysene } { Alizarine.

Pitch 70 % { Used for Roofing and Pavements.
{ Anthracene, 2 per cent.

The preparation of Alizarine from Anthracene:

Anthracene.   Acetic Acid.   Potassic Bichromate.   Anthraquinone.

$$C_{14}H_{10} + C_2H_4O_2 + K_2Cr_2O_7 = C_{14}HO_2 + \text{etc.}$$

Anthraquinone.   Bromine.   Dibromoanthraquinone.

$$C_{14}H_8O_2 + 2\,Br = C_{14}H_6Br_2O_2 + \text{etc.}$$

Dibromoanthraquinone.   Potassic hydrate.   Alizarine.

$$C_{14}H_6Br_2O_2 + {}_2HKO = C_{14}H_8O_4 + \text{etc.}$$

Products Obtained by the Distillation of Coal.

NAME.	FORMULA.	GAS OR VAPOR. SPEC. GRAVITY.	BOILING-POINT. DEGREES.
Atmospheric air.............	....	1.000	
Hydrogen	H	0.069	
Nitrogen.....................	N	0.971	
Oxygen......................	O	1.106	
Ammonia	NH_3	0.590	33
Aqueous vapor..............	H_2O	0.622	100
Carbonic oxide.............	CO	0.967	
Carbonic anhydride......	CO_2	1.529	109 (Fah.)
Cyanogen....................	CN	1.801	
Sulphurous anhydride...	SO_2	2.2112	$+13.09$ (Fah.)
Carbon disulphide.........	CS_2	2.645	47
Marsh Gas Series.			
Methyl hydride	CH_4	0.5596	
Ethyl hydride..............	C_2H_6	1.037	
Propyl hydride.............	C_3H_8	1.522	

I

NAME.	FORMULA.	GAS OR VAPOR. SPEC. GRAVITY.	BOILING-POINT. DEGREES.
Butyl hydride..............	C_4H_{10}	2.005	9
Amyl hydride..............	C_5H_{12}	2.489	30
Hexyl hydride	C_6H_{14}	0.669	65
Octyl hydride..............	C_8H_{18}	0.726	108
Decyl hydride..............	$C_{10}H_{22}$	....	158
Olefiant Gas Series.			
Methylene	CH_2	0.484	39
Ethylene (olefiant gas)...	C_2H_4	0.9784	
Propylene (tritylene).....	C_3H_6	1.452	— 17.8
Butylene....................	C_4H_8	1.936	+ 35
Amylene....................	C_5H_{10}	2.419	55
Caproylene (hexylene)...	C_6H_{12}	2.97	61.3
Œnanthylene..............	C_7H_{14}	3.320	99
Acids.			
Hydrosulphocyanic.......	$H(CN)S$	....	85
Hydrosulphuric..........	H_2S	1.175	
Carbolic (phenol).........	$H(C_6H_5)O$	1.065 (solid)	188
Rosolic....................	$C_{20}H_6O_3$		
Hydrocyanic	HCN	0.7058	26.5
Acetic....................	$C_2H_4O_2$	2.079	120
Alcohols.			
Cresylic alcohol..........	C_7H_8O	....	203
Phlorylic alcohol.	$C_8H_{10}O$		
Benzole Series.			
Benzole...................	C_6H_6	2.695	82
Toluole...................	C_7H_8	3.179	111
Xylole....................	C_8H_{10}	3.179	129
Cumole...................	C_9H_{12}	4.147	148
Cymole...................	$C_{10}H_{14}$	4.632	175
Naphthalene..............	$C_{10}H_8$	4.423	212
Anthracene...............	$C_{14}H_{10}$	6.741	Melts at 213
Chrysene.................	C_6H_4		
Pyrene	$C_{15}H_4$		
		Sp. Gr. $H_2O = 1$	
Aniline...................	$H_2(C_6H_5)N$	1.020	182
Pyridine..................	$(C_6H_5)N$	....	115
Picoline	$(C_6H_7)N$	.09613	134
Lutidine.................	$(C_7H_9)N$	.921	154
Collidine................	$(C_8H_{11})N$	....	170
Parvoline................	$(C_9H_{13})N$	....	188
Coridine................	$(C_{10}H_{15})N$	....	211
Rubidine................	$(C_{11}H_{17})N$	....	230
Viridine................	$(C_{12}H_9)N$	1.017	251
Licoline................	$(C_9H_7)N$	....	235
Lepidine	$(C_{10}H_9)N$	....	260
Cryptidine..............	$(C_{11}H_{11})N$	....	256
Pyrrol..................	$(C_4H_5)N$	....	133

Sign of Death.

The best mode of determining the death of a person is that by Dr. Hugo Magnus, of Breslau. It is simple, conclusive, and easily applied by all. When life ceases and the person is dead, the circulation of the blood positively stops, and is at the end of its action. No matter how inactive the coma or trance, no matter how death-like the lethargy, some circulation will continue, be it ever so sluggish. When it stops once, resuscitation is impossible.

All that is required, therefore, is to tie a string firmly around the finger of the supposed corpse. If there is the least spark of life left, or the blood circulates at all, the whole finger, from the string to the tip, will gradually turn a bluish red, from the enlargement of the veins. Nothing else can be mistaken for this appearance.

Iron.

Iron is produced not only by chemical action in the earth, but also by condensation, etc., in space beyond the influence of our planet. It falls to the earth in the form of meteoric bodies, improperly called meteoric stones: the iron in this form is combined with nickel and other substances. All parts of the world have furnished specimens of these meteoric irons. Some fine specimens are deposited in the Smithsonian Institute, at Washington, similar in form to rings and bombs, and may well be termed Jove's discarded toys, sent to us in remembrance.

Iron is also contained in the human blood: 100 parts of fibrin contain 2.151 of mineral matter and 0.0466 metallic iron; 100 parts of the blood-globules contain 1.325 mineral matter and 0.350 iron; 100 parts of albumen contain 8.715 mineral matter and 0.0863 iron. 100 parts of human blood contain 0.3 fibrin, 7.0 albumen, 12.7 globules, 1.0 mineral matter, and 79.0 water.

The first process in the manufacture of iron is melting the ore in the blast-furnace; here the ore, with coal and a flux of limestone, is piled in and kept burning day and night. The iron, as it becomes melted, flows to the bottom of the furnace, and is drawn off, in a glowing stream, into moulds formed in the sand, which are called pigs, from a fancied resemblance. Prior to the year 1835, wood and charcoal were entirely used in this country, and in that year Frederick W. Geisenhainer, of Schuylkill County, succeeded in producing iron with the hot blast and anthracite coal.

In England, wood and charcoal were entirely used until 1612, when Simon Sturtevant patented the smelting of iron by the consumption of bituminous coal, and it did not come into general use until many years after.

In 1829, Neilson took out a patent in England for the use of the hot blast, and was by this process able to make three times the quantity at less expense. In one single case of infringement of his patent he received a check for damages for three-quarters of a million dollars.

For converting the pig- or cast-iron into wrought, it undergoes the process of " boiling " or puddling, which process was invented by Henry Cort, and patented by him in England, in 1783. For this invention, Cort has been called "the father of the iron trade"—his invention having increased the wealth of England three thousand millions of dollars. In consequence of having expended all that he possessed in experimenting for the grandeur of his country, he died a poor man. The iron being melted in the furnace, the puddler inserts an iron bar through a small opening and stirs the metal until he forms it into a ball. It is then taken out and conveyed to the rotary squeezer. The ball of white-hot iron is placed in the squeezer, and forced with a rotary motion through a spiral passage, the diameter of which is constantly diminishing, which forces out all slag and cinder, and the ball assumes the shape of a short, thick cylinder, called a "bloom." This is again heated in an oven, and conveyed from that to the steam-hammer, where it is pounded into shape and form, or conveyed to the rollers, and by means of them rolled into the various forms, either sheet, round, square, or beams and girders.

The ninth census gives the following information relative to the iron industries of the United States for the year ending June, 1870.

Pig-iron, 386 establishments; 574 blast-furnaces (with a daily capacity of 8,357 tons melted metal), employing 27,554 hands, producing 2,052,821 tons of pig, of the value of $69,640,498.

Foundries, 2,653, employing 51,297 hands, and producing to the value of $99,837,218.

Bloomary forges, 82, employing 2,902 hands, and producing 110,808 tons of blooms, of the value of $2,765,623.

Forges, 103, employing 3,561 hands, and producing to the value of $8,147,669.

Establishments producing bar, rod, and railroad iron, nail plates, etc., 309, employing 44,643 hands, and producing to the value of $120,301,158.

Manufacture of Russia Iron.

The iron for this purpose is refined from pig obtained from magnetic ore, or red and brown hæmatites, by the use of wood and charcoal.

The conversion of pig- into wrought-iron is effected either in the refining fire or the puddling-furnace. The iron must be more granular than fibrous, and must contain sufficient carbon.

The manufacture of iron was carried on in India long before its introduction into Europe, as well as the cutting and polishing of granite. According to a statement by Richard Mallet, the production in India was carried on to such an extent as to rival the production of the largest steam-hammer forges in Europe at the present day.

As an illustration, he mentions that of a wrought-iron pillar or column at the principal gate of the ancient mosque of the Kutub, near Delhi, which is as large as the screw-shaft of a first-class steamer, being sixty feet in length, and surmounted with a capital of elaborate Indian design, carved by chisel in the solid iron. It contains eighty cubic feet of metal, weighs upward of seventeen tons, and contains an inscription dating back fifteen centuries.

Puddling-Furnaces for Iron.

There have been quite a number of improvements made in the construction of the furnaces, some of which are as follows:

Wood and Jackson invented an oscillating or rotary puddling-furnace of cylindrical form, having two skins, between which is a water-space for protecting the sides of the cylinder from heat, and on its inner upper surface a lining of brick, to reduce the amount of space, and so prevent the loss of heat. At each end is a trunnion on which the cylinder turns, and through which the flame from the furnace makes its entrance and exit. These trunnions are also protected by water-chambers. On the top and through the brick lining is an opening, through which the iron is fed to be heated, and through which it is removed when balled, the cylinder being partially rotated for this purpose. Beneath the water-space is located an air-chamber, from which a set

of tuyeres open into the cylinder, and through them a steady
blast is forced through the molten iron, which, combined with
an oscillating motion of the cylinder, it is said, produces a
superior quality of iron.

William Sellers has constructed a puddling-furnace combined with a heat-restoring apparatus, consisting of a series
of tubes through which the waste heat passes, arranged in a
similar manner to those in the furnace of Gorman and Paton,
but differing in being divided into two sets — one of which
heats gas, and the other air — so arranged that the gas and
air, after being heated, meet in opposing columns so as to be
thoroughly mixed before combustion, at the mouth of the
rotary furnace, where the mixture becomes ignited, and passing through the furnace, acts on the metal therein, and then
enters the regenerating tubes, leaving there the most of the
waste heat to be again taken up by the incoming air and gas.
To remedy a difficulty experienced in starting this class of
furnace, owing to the cold air contained in the passages, a
flue is provided in which fire may be introduced to heat the
air, and so induce a current.

Improved Cupola Furnace.

Smelting iron in a cupola furnace appears to most people,
who see it daily done at every foundry, the simplest thing in
the world; it is, however, not so, if due regard is taken to
economy and good quality in casting. In a common cylindrical cupola, three essential parts may be distinguished.
The upper half or body of the furnace prepares the pig-iron
and lime, which, together with coke, are thrown in at the top
for smelting in the middle part or crucible, which is somewhat narrower, and provided with numerous nozzles for the
introduction of blast, whence the molten iron, together with
slag, runs down to the lower part, or hearth, where it collects
until it is tapped.

When such a furnace is to be started, it is filled to about two-thirds with coke, and one-third with coke and iron; fire is then introduced and the blast turned on, when the molten iron collects in the hearth, and replaces the coke of the same. Here it necessarily takes up impurities from the coke, and impregnates the latter so much that it cannot be destroyed by the blast, and when the iron is tapped, masses of coke and half-melted iron, which are not any longer supported, tumble down in the hearth, where they are imperfectly burnt or melted, and cause the iron which collects there to become cold and sticky. These irregularities take place after every tap, and it generally happens that iron which was at first fluid and gray, suddenly becomes thick and white, and unsuitable for the castings intended. In order to avoid this, Henry Krigar, of Berlin, constructs his cupola so that the lower part, or hearth, is not below the crucible, but by its side, and connected with it by a slanting canal, which is about 3 inches high, 6 to 8 inches long, and as wide as the cupola. This arrangement prevents any coke or half-melted iron from falling down in the hearth, which is only accessible to melted iron and slag, and forms for them a kind of sump or receiver, which in no way interferes with the regular working of the two upper parts of the cupola. This very simple construction has proved highly successful, and its great advantages are a saving of fuel, a uniformly hot and liquid iron, and an increased yield per diem, as the regular smelting operation is never interrupted. Krigar's cupola can, therefore, be recommended not only for foundries, but also for Bessemer works.

Henderson's processes consist in the use of fluorine, in conjunction with oxygen, which, acting on the molten cast-iron, in a few minutes almost completely purifies it. They are applied to wrought-iron as follows: Fluorspar and oxides of iron are ground to a fine powder, mixed thoroughly, and thrown into the ordinary puddling-furnace. The molten cast-iron to be acted on is then poured over the spar and oxides, which remain on the bottom of the furnace; the furnace-door is closed, and the iron allowed to boil for about half an hour; the rabble is then worked for about ten minutes, and the metal is balled up in the usual way. The whole time occupied by one charge, with ordinary gray-forge pig-iron, being about one hour. The mode in which the purification of the iron is effected appears to be as follows: After the cast-iron has been poured in and the furnace closed, the heat of the metal melts the fluorspar and oxides, which, combining with the silicon, sulphur, and phosphorus, and part of the carbon, removes them from the cast-iron in the form of vapor and slag, and this so effectually that, by the time ordinary gray-iron has had the amount of carbon in it reduced to $2\frac{1}{2}$ per. cent., common brands of pig-iron are rendered as free from impurity as good Bessemer steel.

The removal of impurities by the Henderson processes enables iron of a very high quality to be made from cinder-pig and other common brands of cast-iron. Although the tensile strength of the iron is very great, it is chiefly remarkable for its toughness, as will be seen from the high elongation. Great elongation is one of the most valuable and peculiar properties of the iron made by the Henderson processes. For instance, plates made in England, at Bowling Iron Works, Bradford, from Yorkshire cinder-pig iron, gave more than double the elongation of Lowmoor, Bowling and Farnley.

The tests were made by Mr. Kirkaldy, and the following is a summary of them:

	Ultimate tensile strength, tons.	Contraction of area, per cent.	Elongation, per cent.
Henderson's cinder boiler-plate, Mean of length and crossway.	24.73	32.4	23.7
Bowling, Lowmoor and Farnley, Mean of length and crossway.	23.60	16.6	11.27
	1.13	15.8	12.43

12 *

Difference in favor of Henderson's cinder-iron, } 4.78 per cent.
tensile strength.. }

Difference in favor of Henderson's cinder-iron, } 100 per cent.
contraction of area.................................. }

Difference in favor of Henderson's cinder-iron, } 110 per cent.
elongation.. }

Steel manufactured from wrought-iron made by these processes from common Scotch pig-iron gave a tensile strength equal to steel made from the best Swedish iron; when made into tools, it also stood wear equally well.

Ira Hersey, in his process for converting iron into steel, has constructed a furnace in which the balls from a bloomary are transferred to an inclined hearth-floor in which the proper chemicals for carburizing — manganese, prussiate of potash, charcoal, and salt—are introduced in thin sheet-iron cans. The metal as fast as melted combines with the chemicals and runs into a receiving basin at the bottom, from whence the steel so formed may be drawn to form ingots.

Barron's process for converting cast-iron tools into steel. Tools which are to be prepared by this process are first made of cast-iron, after which they are introduced into a revolving drum, and all roughness worn off by attrition. The smooth irons are then packed in layers in iron boxes, where they are closely imbedded in clay, and subjected to the action of oxide of iron and certain chemicals, by which the iron is decarburized. In these boxes the iron tools are subjected to an annealing heat, which is maintained from three to six days. They are subsequently placed in a retort capable of containing about a ton of the tools, into which the vapor of gasolene and other carbonaceous materials are passed. In a few minutes the iron is converted or transformed into steel, when they are ready to be tempered, and ground and polished for the market.

A. H. Siegfried's process for tempering steel by heating it
to a cherry-red in a fire purified by throwing in salt, then
covering the steel with the same substance, and, while sub-
jecting it to this treatment, working it nearly to the finished
form. A compound of the following ingredients is then
substituted for the salt: one part each of salt, sulphate of
copper, sal ammonia, and sal soda, and a half part of nitrate
of potash; and the steel is alternately heated and treated
with this until it is refined and brought to the finished form.
It is then slowly heated to a cherry-red, and plunged into a
bath of the following ingredients: alum, sal soda, and sulphate of copper, of each one and a half ounces; nitrate of
potassa, one ounce, and salt, six ounces, dissolved in a gallon of rain-water.

Ancient Granite Works in the East Indian Empire.

The art of cutting or carving in granite has never been
equalled or carried to as great perfection as on the conti-
nent of India. At Chillambaram, in the Carnatic, and on
the Coromandel coast, is a congeries of temples representing
the sacred Mount of Meru. Here are seven lofty walls, one
within the other, round the central quadrangle, and as many
pyramidal gateways in the midst of each side, which form
the limbs of a vast cross, consisting altogether of twenty-
eight pyramids. There are, consequently, fourteen in a line,
which extend a mile in one direction. From the nave of one
of the principal structures there hang, on the tops of four
buttresses, festoons of chains, in length about 548 feet; each
garland, consisting of twenty links, is made of one piece of
granite 60 feet long; the links themselves are monstrous
rings, 32 inches in circumference, and polished as smooth as
glass. Compared with the cave temples of the southern Car-

natic, and the monolith temples of granite at Mahabali-
pooram, those in Egypt sink into insignificance.

The stone of which the cave temple of Elephanta is composed resembles porphyry, and the elaborate tracery and
sculptures with which the columns at the entrance, and also
of the interior, are decorated, are exquisitely delicate; but
ignorance, superstition, and blind fanaticism have committed
strange and barbarous havoc. The area occupied by this
temple is nearly 130 feet deep and about 133 broad, divided
into nine aisles formed of twenty-six pillars, of which eight
are broken away altogether, and most of the remainder are
much injured.

KEYLAS.—This cave is excavated from the solid rock, and
is the most elaborately designed and artistically enriched of
the caves of Ellora. The height of the outer gateway of
Keylas is fourteen feet, opening to a passage with apartments
on either side. Over the doorway is the Nogara Khana, or
music gallery, the floor of which forms the roof of a passage
leading from the entrance to the excavated area within.
Entering upon the latter, which is a wide expanse of level
ground, formed by cutting down through the solid rock of
the hill, an immense temple of a complex pyramidal form
presents itself, connected with the gateway by a bridge, constructed by leaving a portion of the rock during the progress
of the excavations.

In front of the structure, and between the gateway and
the temple, are the obelisks of Keylas, placed one on each
side a pagoda or shrine, dedicated to the sacred bull Nundee. These obelisks are of a quadrangular form, eleven
feet square, sculptured in a great variety of devices, all of
which are elaborately finished; their height is about forty-
one feet, and they are surmounted by the remains of some
animal, supposed to have been a lion, which, though not
an object of Brahmanic veneration, occurs very frequently

among the decorations of the cave temples. Approaching
the entrance to the temple is a colossal figure of Bhawani,
supported by a lotus, having on each side an elephant, whose
trunks form a canopy over the head of the goddess.

On each side of the passage, from the inner entrance, are
recesses of great depth and proportions, in one of which,
resting upon a solid square mass, is the bull Nundee, su-
perbly decorated with ornaments and rich tracery. Beyond
this, on the opposite side, is a similar recess, in which is a
sitting figure representing Boodh, surrounded by attendants,
and near the end of the passage, where the body of the great
temple commences, is a sitting figure of Guttordhirj (one of
the incarnations of Siva), with his ten hands variously occupied. Turning to the right, the walls of the structure are
covered by sculptures representing the battles of Ram and
Rouon, in which the achievements of the monkey-god, Hu-
mayun, are conspicuously displayed. Pursuing the story de-
picted by these sculptures to the end, the extremity discloses
the entrances into three temples. Returning to the entrance on the left, the history of the war of the gods is continued in sculpture. The whole length of the substructure
appears to be supported on the backs of animals, such as
elephants, lions, horses, etc., which project from the base of
the piers in the surrounding walls, and give the superin-
cumbent mass an air of lightness and movability.

This extraordinary structure is in every portion of the
exterior, as well as the interior, carved into columns, pilas-
ters, friezes, and pediments embellished with the represen-
tation of men and animals, singly or in groups. The gal-
leries contain sculptured history, and forty-two gigantic
figures of gods and goddesses. A portion of the chambers
is richly and lavishly embellished, one containing groups of
female figures so exquisitely proportioned and sculptured,
that even Grecian art has scarcely surpassed the beauty of

the workmanship. Pen and pencil can afford very ineffectual aid in a field so vast and unparalleled as that of the Keylas of Ellora.

Their sacred character has been lost in the obscurity of unknown ages, and the projectors must have possessed intellectual and imaginative gifts of extraordinary power. The rock from which this temple is wrought is a hard, red granite, and from every peak and pinnacle of the sacred mountain the eye roams over scenes of romantic beauty and marvellous grandeur.

According to the Brahminical account of the origin of these excavations, 7,894 years have elapsed since they were commenced, as a work of pious gratitude, by Ecloo Rajah, son of Peshpout of Ellichpore, when 3,000 years of the Dwarpa Yoag were unaccomplished, which, added to the 4,894 years of the present, or "Kal Yoag," completes the full number, 7,894.

The interior of the cave temple of Elephanta is covered with sculpture from the entrance to the farthest recess of the excavation, and mystic forms meet the eye in every direction. A colossal triple-headed bust occupies a vast recess at the extremity of the central aisle of the temple. The dimensions of this relic are, from the bottom of the bust of the central figure to the top of the cap on its head, eighteen feet; the principal face is five feet in length, and the width, from the front of the ear to the middle of the nose, is three feet four inches. The width of the whole bust is twenty feet. A cap surmounts the head of the central figure, once richly decorated with superb jewels. Around the neck of the same figure was formerly suspended a broad collar, composed of precious stones and pearls, long since appropriated to a more useful purpose than the decoration of a block of carved stone in the bowels of a mountain.

The whole of this singular triad is hewn out of the solid

rock, which is a coarse-grained, dark-gray basaltic formation, called trachyete, and it occupies a recess cut in the
rock to the depth of thirteen feet. On each side of the
niche is sculptured a gigantic human figure, having in one
hand an attribute of the Deity, and with the other resting
upon a dwarf-like figure standing by its side. Niches, or
recesses of large dimensions, and crowded with sculpture,
appear on either side.

In that on the right-hand side is a colossal figure, apparently a female, but with one breast only. This figure has
four arms: the foremost right hand rests on the head of a
bull; the other grasps a cobra-de-capello; a circular shield is
borne on the inner left hand. On the right of this female
is a male figure, bearing a pronged instrument representing a trident; on the left, a female bears a sceptre. Near
the principal figure is an elephant surmounted by a beautiful youth, and above the latter is a figure with four heads
supported by birds. Opposite to these is a male figure
with four arms, sitting on the shoulders of one in an erect
posture, who has a sceptre in one of the hands; and at the
upper part of the back of the recess are numerous small
sculptured figures, in various attitudes and dress, supported
by clouds.

Turning to the niche on the left is a statue of a male,
seventeen feet in height, having four arms; to the left of this
a female fifteen feet in height. The countenance of this statue
is sweetly feminine, and expressive of gentleness and amia-
bility. The head-gear of the small figure bears a resemblance to the wigs worn by modern judges.

Various conjectures have been hazarded by the learned
as to the origin and purposes of these extraordinary cavern
temples. The following explanation of some of the sculptures
is from a paper preserved among the collection of the Asiatic
Society of Bengal.

The triple-headed colossal bust described in this document is the personification of the great attributes of that being for whom the ancients, as well as the Hindoos of the present day, have entertained the most profound veneration. The middle head represents Brahma, or the creative power; that on the left, the same deity as Vishnu, or the preserver; that on the right, Siva, or the destructive power. The bull couchant at the feet of one of the deities symbolizes an attribute of Siva, under his name of Iswara. The beautiful youth on the elephant represents Cama, the Hindoo god of love.

The terrific figure with eight arms represents the destroyer Siva in action. The distant scene with small figures, expressive of pain and distress, denotes the sufferings of those sentenced by Brahma to the place of torment; and it is considered that the whole was dedicated to the worship of the god Siva, and to the mysteries of his cruel and impure ritual.

THE END.